Engineering Design III

Engineering Design III

M. H. A. Kempster

C. Eng., M. I. Mech. E., M. R. Ae. S., M. I. Prod. E.
Senior Lecturer in Mechanical and Production Technology
Rolls-Royce Technical College, Bristol

HODDER AND STOUGHTON
LONDON SYDNEY AUCKLAND TORONTO

First printed 1984

Typeset in 11/12pt Univers (monophoto) by Macmillan India Ltd., Bangalore

Printed in Great Britain for
Hodder and Stoughton Educational,
a division of Hodder and Stoughton Ltd.,
Mill Road, Dunton Green, Sevenoaks, Kent TN13 2YD,
by J. W. Arrowsmith Ltd., Bristol BS3 2NT

Contents

Acknowledgement

Cover: gears provided by Ketlon (UK) Ltd.

Preface

Engineering design is a problem-solving activity which, like all such activities, requires the precise definition of the problem. In this book, which is suitable for students of engineering design at BTEC level III and above, three aspects are emphasised. They are the definition of the design requirements, the identification of the problem areas associated with the principal manufacturing processes and the methods whereby these problems can be overcome.

Engineering design requires a sound understanding of several disciplines. A knowledge of mechanical science and mathematics allows the problem to be solved using basic principles and the necessary calculations to be made. A knowledge of materials and manufacture means that the construction can be considered. Engineering drawing is a necessary discipline because it is the principal means of the communication of ideas. The reader is reminded that a subject can only be learned by active participation, and that this is particularly true of any subject involving design. He or she should therefore examine as many products as possible and consider both the good and bad features of each, and then follow this up with the preparation of as many original designs as possible, which should be critically examined by a competent designer.

The writer wishes to acknowledge the work done by his wife, who checked the original manuscript and generally assisted in the preparation of this book.

M. H. A. Kempster
Bristol

1 Planned Design

Design is a problem-solving activity and, as with all such activities, the most difficult part is the precise identification of the problem.

The initial specification of a product that is to be marketed is usually produced by a company's sales division acting under the direction of a strategic planning committee of the board of directors. This specification may be based upon the results of a market survey combined with a study of possible competitors, and it usually includes the possible selling price of the proposed product. This specification would be studied by the strategic planning committee which would take into account the design, development and manufacturing capabilities, as well as the future development of the company, before passing the proposal over to the design division.

Alternatively the initial specification could come from a potential customer who would then invite tenders. A typical example is in the aerospace industry in which an airline operator or a government department usually states the requirements (for example, the speed, range and payload) of a proposed aircraft. Manufacturing companies then compete for the contract to design and produce the airframe and engines to satisfy the requirements.

Equipment such as manufacturing aids and test plant, which is used within a company, is usually studied on an inter-departmental basis. Thus the manufacturing area and the test engineers assume the role of the customer. Whatever the system whereby the initial specification is produced, the method used to translate it into the final design always follows a basic pattern which can be summarised in five steps.

1 Identification of the problem.
2 Analysis of the problem.

3 Design synthesis.
4 Design evaluation.
5 Development oт the selected design.

1.1 Identification of the problem

The total problem to be solved usually comprises several problems which are not necessarily directly related and which must be identified at an early stage in the design process. These problems can be classified as problems associated with:
(a) function;
(b) construction;
(c) appearance.

(a) Problems associated with function

The problems associated with function can be identified as the satisfaction of both the primary function (or functions) and the secondary function (or functions). The secondary functions are those which must be fulfilled to enable the primary functions, to be performed.

The definition of the primary function, or functions, must include the precise duty of the product or device, the accuracy with which it must perform that duty and the circumstances in which the product will be used. These factors may limit the method of operation. Other factors must be taken into account, such as the required degree of reliability of the product and the frequency of its use. Although the customer usually considers reliability to be of great importance, it usually increases the cost of a product; often it may only be achieved at a reasonable cost if a low level of performance is acceptable. When a device is intended for use only in emergency it produces special problems because, although it is hoped that its use will never be necessary, it must be completely reliable when needed.

The secondary function should also be identified at this point because the extent to which the primary function can be performed is usually controlled by the secondary functions. This can be illustrated by the apparently simple problem of providing means of keeping a person dry when shopping in wet weather. To fully satisfy this primary function the device should completely envelop the user; to satisfy the function as a whole, however, several secondary functions must also be satisfied. The device must be stored and be easily transported when not fulfilling its primary function. It must be easily and rapidly prepared for use in the event of a storm, and it must not be excessively inconvenient to use (particularly when other people are using similar devices).

The device must also be rapidly dismantled, transported or given temporary storage at the destination or when the storm has passed. The need to satisfy the secondary functions results in the device being an umbrella. This device only partly satisfies the primary function, but it satisfies most of the secondary functions.

The safety of the user and of people in the vicinity of the product, together with the protection of the mechanism in the event of its being overloaded, is usually considered to be associated with the primary function. The need to ensure that the product is safe may dictate the method of its operation and, in extreme cases, make the product unsuitable for sale to and use by the general public.

(b) Problems associated with construction

The problems associated with construction are related to the function of the product and its maintenance. They are also related to the manufacturing method that suits the available and appropriate manufacturing equipment, the delivery date, the rate and volume of manufacture, and the allowable cost. Further problems related to the material and shape will also be revealed at the analysis and synthesis stages.

(c) Problems associated with appearance

Many of these problems are directly related to function and to construction, but ergonomic considerations (how easy the product is to use) and the need to satisfy 'selling points' may also affect the appearance of the product.

(d) Identification of the problems by a self-interrogation system

A basic *self-interrogation* system is presented in this section, but readers may well develop their own systems to satisfy their personalities or their particular industries and use them at all of the design stages. Many of the questions will not be relevant to a specific problem, but their consideration will ensure that all aspects are considered. As in all design work this stage must be approached with an open mind because any attempt to pre-judge the issue will defeat the object of the study. It is convenient to relate the self interrogation to function, construction and appearance.

(i) Function

- What is the precise duty of the device or product?
- To what accuracy must the duty be performed?
- How fast must the device operate?

- Under what conditions must the device operate?
- Are there any limitations regarding the method of operation (for example, is electricity unsuitable or unsafe)?
- Will the device be used frequently?
- How reliable must the device be?
- Are there any safety, legal or insurance requirements that must be met?

(ii) Construction

- How strong must the device be?
- Does the function demand special construction (for example, to hold water or air under high pressure)?
- How large will the device be?
- Will the complexity of the device demand a special construction system?
- Will the environment in which the device is to be used affect the choice of material? (For example will creep, corrosion or high temperatures cause problems?)
- Are there any aspects of assembly or maintenance that affect construction?
- Are there any safety, legal or insurance regulations that limit the construction method?
- What will be the total quantity to be produced, and what will be the rate of production?
- Is there a delivery promise that will affect the method of construction?
- Are there any construction limitations imposed by the availability of equipment or plant to manufacture the device?

(iii) Appearance

- Are colour, shape or texture 'selling points'?
- Are colour, shape or texture associated with ergonomic considerations? (For example, the colour of a screwdriver handle will affect the speed with which it is located on a bench, its shape will affect the user's posture and its texture will affect the efficiency with which it is gripped.)
- Is colour important as a means of identification?
- Does shape or texture improve the product or extend its life? (For example, the shape of a car body may affect the performance of the car, and the texture of a part its fatigue life.)
- Does the method of manufacture determine the shape and appearance of the product?

1.2 Analysis of the problem

When the problem has been identified, it can be analysed. Analysis determines more precisely the requirements of the product, identifies any features of the initial specification that may be impractical or unsuitable and indicates the extent to which the target selling price may be achieved. The strategic planning committee would be informed of such findings and a decision made regarding any revision to the initial specification.

1.3 Design synthesis

When the design requirements have been established, the methods whereby they may be satisfied can be examined by producing design studies. At this third stage of the design process, initial calculations can be made to determine the required power, the size and shape of the major components and the material from which they will be made. Usually there will be several solutions to the problem, all of which should be considered even if they are immediately rejected. Good solutions are usually generated by producing a number of solutions, combining their best features to produce further solutions and, in turn, producing further solutions from them. These solutions must be evaluated and eventually the most suitable one is developed.

1.4 Design evaluation

The process of design evaluation occurs at the design synthesis stage and when the selected design is being developed. Individual solutions should be evaluated by a continuation of the self-interrogation system in which the design is subjected to appraisal against the function, construction and appearance requirements already identified. Self interrogation should proceed on the following lines (the precise questions will depend upon the requirements already identified).

- Will the device, as designed, fulfil its primary function?
- Can it operate at the speed that is required?
- Will the device function correctly when in service conditions?

As a result of the examination several solutions may be rejected. If several solutions, each with some good features and some poor features, need to be considered, then the solution that is the best compromise must be identified by using an assessment system in which the solutions are either compared with each other or with an expected 'profile'.

(a) Comparison of solutions with each other

In this system the most important requirements are listed and each solution is rated with respect to each requirement. The usual rating system is a 0 to 9 point system, in which 0 represents unsuitable and 9 represents excellent. The solution with the greatest number of points is the best all-round solution.

This system is illustrated in table 1.1 by considering four solutions each of which, to some extent, satisfies the four requirements listed.

Table 1.1

Requirement	Ideal Solution	Solution 1	Solution 2	Solution 3	Solution 4
Target cost	9	6	5	8	2
Appearance	9	5	8	4	7
Strength	9	8	4	9	6
Suitability for service conditions	9	8	4	8	7
Total points	36	27	21	29	22

Using this system, solution three, which scores 29 points, is the one which is the best all-round solution.

(b) Comparison with an expected profile

In this system (as in the previous one) the most important requirements are listed and each solution is rated with respect to each requirement. But, instead of comparing them with an ideal solution that scores maximum points, an expected profile is produced which gives a higher rating to the most important requirements.

Using the previous example the rating of the four solutions remains the same, but the rating of the expected solution, the profile, is as shown.

Requirement	Profile
Target cost	8
Appearance	5
Strength	7
Suitability for service conditions	8
Total points	28

These ratings can be presented in the form of a diagram as shown in fig. 1.1. In this example, solution three satisfies the cost, strength and service requirements of the expected profile but falls below its appearance requirement. Provided that its appearance is acceptable it is the best all-round solution when this system of evaluation is used.

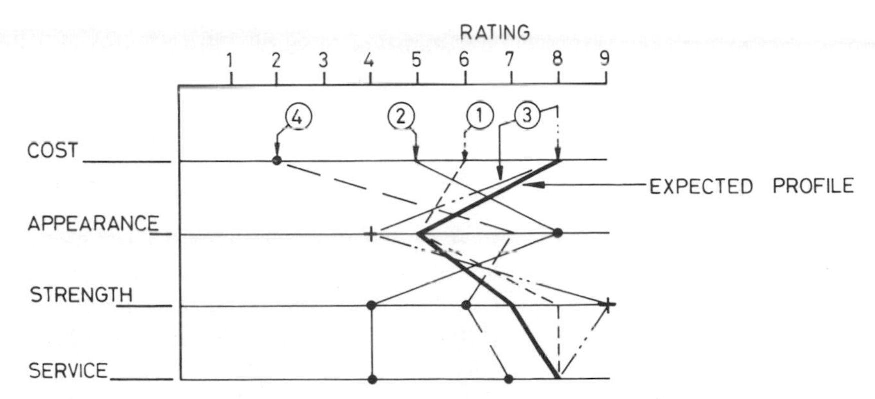

Fig. 1.1 Rating diagram

(c) Other applications of evaluation systems

The evaluation systems described in this section can also be used to evaluate specific aspects of the design. (The design analysis in section 1.7(a) uses such a system to evaluate the source of illumination for a table lamp.) Methods of manufacture, material for a specific component etc. can also be evaluated in the same way.

1.5 Design development

Several versions of the selected design are developed at this stage. The versions can be evaluated by self interrogation and compared using one of the evaluation systems described in the previous section. At the design development stage a greater emphasis must be placed upon design for manufacture and its effect upon the cost. The manufacturing and costing departments should be consulted so that the design selected for prototype development is suitable for manufacture and will be of an acceptable selling price.

It is usual to run a series of tests on the prototype to enable its shortcomings to be identified and modifications to be made before the production drawings are prepared. As each change is made the manufacturing and costing departments are again consulted. If these changes alter the basic requirements the sales division is also consulted.

1.6 Information and standards

The designer should take advantage of all the information that is available. The journals of the engineering institutions, and similar learned bodies, and the publications of the various materials development and manufacturing industries associations are invaluable sources of information.

Wherever possible, the design should include proprietary components in order to minimise the cost of the product. For the same reason national standards (such as British Standards) should be used for components, features such as screw threads and for materials. Standards are necessary, both when designing components and when specifying materials etc.

A well-organised design office usually includes a standards section with the specific duty of compiling and circulating in-house standards and procedures. Whenever possible the standards are produced either by abstracting the relevant parts of national standards that suit the size and the required quality of the products of the company or by using data relating to proprietary components that are used. In-house standards may, however, be produced as a result of development work done in the company. Procedures may also be specified to suit legal requirements, customer's needs or as a result of manufacturing development.

A company often maintains a stock of material or has a ready supply of material from a stockist. A list of preferred materials, their properties and, where appropriate, the preferred sizes (such as the diameter of bar stock) is often prepared.

The technical library usually maintains carefully indexed sections that hold abstracts from technical reports, lectures etc. and also details of proprietary components that may be used. Large organisations often employ a qualified librarian whose duties include the search for, and the procurement of, specialised information.

1.7 Typical design studies

The following examples will illustrate the method used to produce initial designs and to evaluate alternative designs.

(a) Design study of a table lamp

An adjustable table lamp is required to illuminate an area of approximately 0.7 m^2 for both domestic and office use. It is to be marketed at a maximum basic price of £25 and large-quantity manufacture is required.

Examination of the above specification indicates that the lamp should be fully adjustable in both the vertical and horizontal directions, be safe and stable, give a steady and fairly high level of illumination, be reasonably reliable (although reliability is not associated with safety) and be of a shape and colour that is functional (for office use) yet is acceptable in the home. The marketing outlet will include mail order and cash-and-carry warehouses, and so the lamp must be easily packed. If it is to be dismantled for packing it must be easily reassembled. Some of these requirements are illustrated in fig. 1.2.

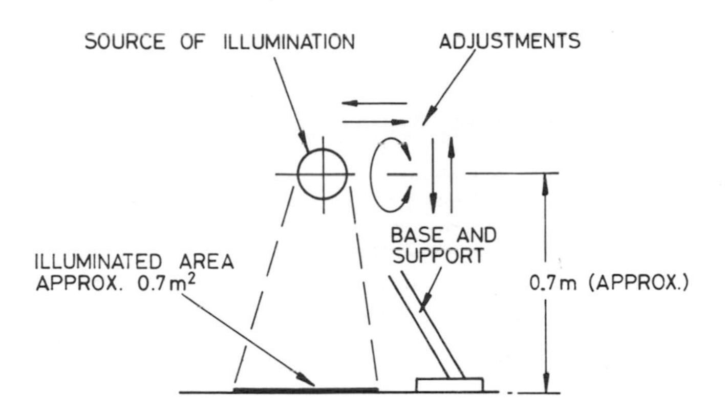

Fig. 1.2 Specification of table lamp

(i) Examination of Costs. A lamp with, for example, a recommended retail price of £25 will need to be sold by the manufacturer to a wholesaler for about £13—the difference between £25 and £13 being taken up by the runnng costs of the wholesaler and retailer, transport costs and profits. If the manufacturer supplies the lamp directly to the retailer it will sell for about £18, and this may be reduced to about £15 when the manufacturer has a direct retail outlet. In all cases the actual price depends upon the running costs, profit margins and quantities involved.

In this example the £13 manufacturer's price includes the cost of the fittings (switch, bulb-socket, flex etc.), the cost of the material, direct labour costs, the cost of sales promotion, packaging, transport etc. and indirect labour costs (technical staff, supervision, management etc.). The rates on premises, depreciation allowance and the insurance associated with premises and equipment as well as taxes on profits and profits payed to the shareholders must all be considered before the price is fixed.

(ii) Method of illuminations. Artificial light for office or domestic use may be obtained from a candle, an oil and wick flame, a gas flame or an electric light. To satisfy the requirements of a steady and fairly high level of illumination, and safety, the only suitable source is an electric light using either a filament bulb or a discharge tube.

The source of electricity may be the mains supply or a dry battery (the latter may be expendable or rechargeable). Each of these sources have both good and bad features, and the best all-round source can be identified by using a rating system with respect to the seven considerations shown in table 1.2. The numbers range from 0 to 9. Convenience relates to freedom from the effects of industrial action, freedom from recharging time and from battery purchase time.

Table 1.2

Consideration	Mains operated	Expendable battery operated	Rechargeable battery operated
Cost of the lamp	8	5	3
Running cost	8	2	4
Convenience	8	6	3
Uniformity of level of illumination	8	4	4
Freedom of location	0	8	8
Safety (flex hazard)	3	8	8
Shape of lamp (for battery)	8	5	3
Total points	43	38	33

From this study it is decided to design a lamp that is operated from the mains. As already stated, electricity can be used in conjunction with a filament bulb or a discharge tube. A filament bulb does not require special fittings and can fit into a shade that suits the home: a discharge tube requires special fittings and needs a shade that does not readily suit the home. A filament bulb, therefore, is more suitable for the lamp under consideration.

(iii) Bulb support and base. The fundamental requirements are that the support system allow the necessary adjustment and that the base is heavy enough to produce the required stability. Three acceptable support systems are shown in fig. 1.3.

System A allows adjustment in the horizontal and vertical planes in addition to the rotation of the shade and socket unit, but it is unsuitable for rapid, frequent adjustment because the block must be held whilst it is locked in position on the vertical rod. It is not particularly neat because the flex must either be clipped to the rod and introduced into the tube (which makes vertical adjustment awkward) or introduced directly into the tube (which makes adjustment less awkward but is hazardous because the loose flex may then be snagged).

System B provides full adjustment and is neat because the flex can be passed from the base through the tube and the coil, but it is not completely suitable for frequent and rapid adjustment because moving it is a two-handed task.

System C provides the required adjustment and, if the links are hollow, the flex can be hidden. The links can be locked in position at the joints by a hand nut or, if frequent and rapid adjustment is required, the joints can be made fairly stiff and retained in the

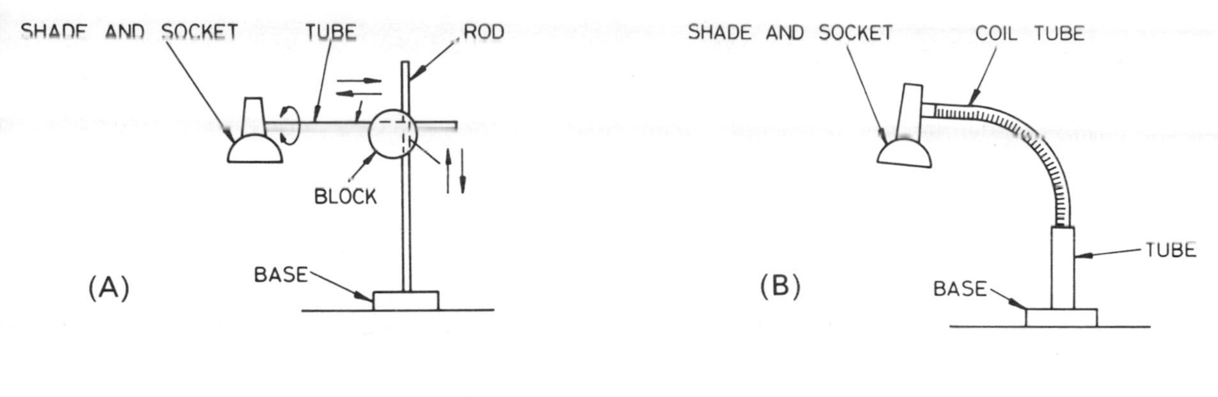

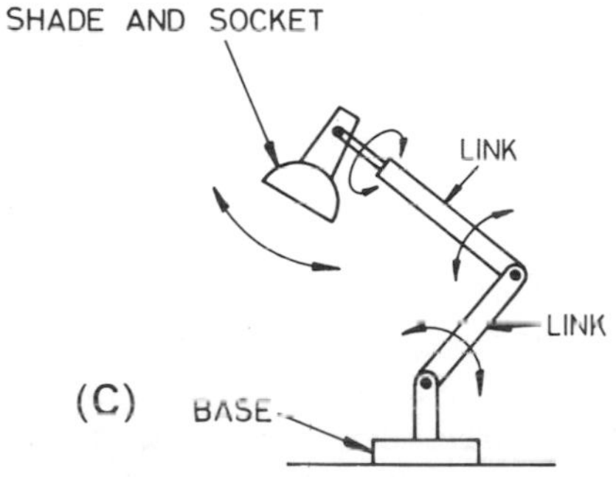

Fig. 1.3 Support systems

required position by a spring system (although this will increase the cost).

The base could be made of wood (with a suitable ballast) but it would not be in keeping with the lamp as a whole. A sheet-metal base (filled with a suitable ballast) or a cast-metal base would be more satisfactory.

(iv) Electrical fittings. A mains-operated lamp requires a socket for the bulb, a length of flex, a switch and a plug. The customer usually expects to supply a plug in addition to a bulb, but the provision of a generous length of flex helps to keep the customer satisfied.

Five basic systems are illustrated in fig. 1.4.

System A does not include its own switch—but it is now usual for wall points to include a switch. This is the least expensive system, but it is inconvenient to use unless the lamp is placed very near to the mains point.

System B includes a switch between the lamp and the wall point. It is more convenient to use than is system A, but it is more expensive because of the cost of the switch and of its installation. Also it is somewhat untidy.

System C includes a switch that is mounted on the base. This is the most convenient system to use, and costs about the same as system B, but requires the base to be designed to take the switch.

System D includes a switch between the base and the bulb

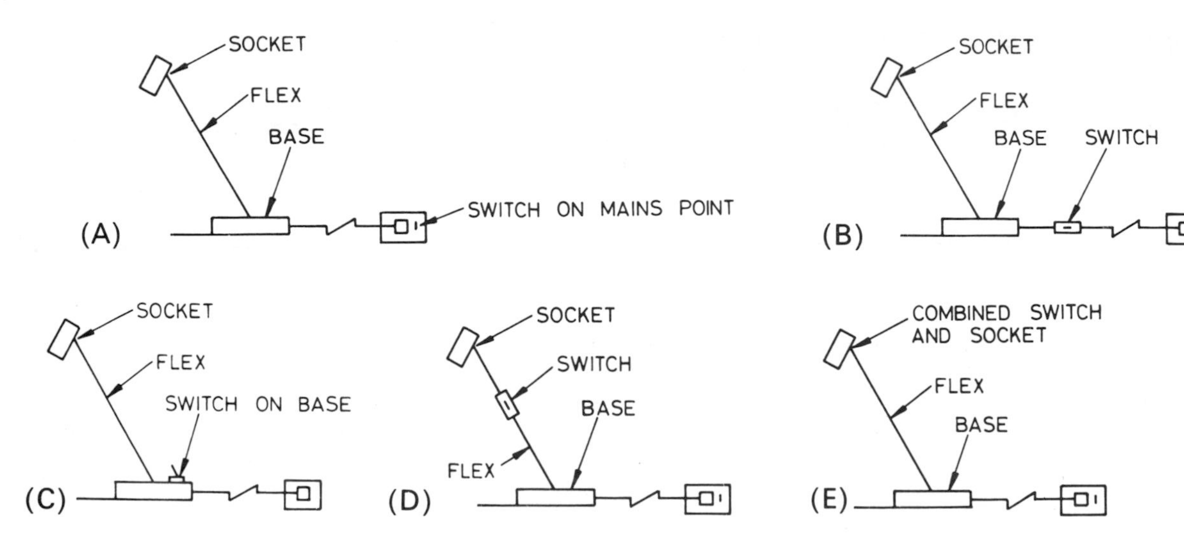

Fig. 1.4 Basic electrical fittings systems

socket. It is basically the same as system B but is more convenient to use, although somewhat untidy.

System E has a combined socket and switch unit which is less costly to purchase and to install than a separate socket and switch. It is convenient to use and is a suitable alternative to system C, but it is not as neat.

System C satisfies the requirements of the lamp under consideration, but system E is an acceptable alternative if it becomes necessary to reduce costs.

(v) Shade. In this example the function of the shade is:

- to direct the light to the area to be illuminated;
- to prevent glare;
- to make the lamp attractive.

To satisfy the first function the shade must be of a suitable shape and be reasonably reflective. If the light is to be in the form of a parallel beam the shade must be a paraboloid with the bulb filament at the focal point—but in this example the beam will need to be of conical form and so a 'modified paraboloid' is required. The inside of the shade can be of polished metal or be painted white to give good reflective qualities.

To satisfy the second function the shade must be designed so that the bulb cannot be seen by the user when the lamp is on.

To satisfy the third function the shape must be of an acceptable shape and colour. The required beam form does not impose special limitations upon the shape, but the colour needs careful consideration (see below).

When designing the shade it must be remembered that the bulb will be hot when in use, and so the shade must be made of a suitable material. It must be large enough for air to circulate between it and the bulb, and it must have ventilation holes. The shade must also be designed so that the space between it and the bulb is large enough, and of a suitable shape, to enable the user to fit and to remove the bulb from its socket.

The shade should, therefore, be made from sheet metal by presswork or by spinning, be of a suitable shape and size, and be polished and/or painted.

(vi) Colour. The colour of the lamp must be acceptable for both the office and the home. It may be possible to produce the lamps in a range of colours determined by market research, but if it is policy to produce the lamps in one colour then an acceptable colour, such as white, should be used.

(vii) Assembly. The method of assembly must be carefully considered at the design stage because assembly will be an expensive cost element. The assembly time can be reduced by using 'snap together' parts, pins and roll pins, and rapid adhesives.

(viii) Summary. This design analysis indicates that the full design specification is as follows.

The lamp should be mains operated and use a filament bulb. The support should be either a balanced link or a flexible coil (depending upon the required market cost) with a heavy base that includes a switch. The shade should to be of sheet metal and include the bulb-socket, and the lamp should either be white or suit the current fashion. A generous length of flex should be supplied but not a plug or a bulb.

(b) Design study of a demonstration rig

Fig. 1.5 shows a typical plate clamp. For effective clamping L_H must be equal to, or larger than, L_w and the purpose of the demonstration rig is to convince the student that this is so. The function of the spring in fig. 1.5 is to support the clamp plate during the loading and unloading of the workpiece, and the function of the lower washer is to prevent the spring from entering the slot or hole in the clamp plate. As these components are not related to the principle to be demonstrated they may be omitted from the rig. The nut may be a hexagonal nut, tightened with a spanner, or a hand nut, depending upon the clamping force that is necessary. A hand nut will be more appropriate in the demonstration.

The basic requirement of a demonstration rig or a visual aid is that it should resemble the actual equipment or situation and not be so involved as to distract from the demonstration, but some licence is usually necessary.

In this case, a range of L_H:L_w ratios must be produced so that the full effect can be demonstrated. This can be done either by providing a set of clamps or by changing the position of the stud in relation to the heel pin and the workpiece. Although normally the stud is fixed, an adjustable stud (in the form of a square-headed bolt located in a T slot) is a better arrangement for a demonstration rig. By marking the base to indicate the $L_H:L_W$ ratio for a number of stud positions the time required for the demonstration is reduced.

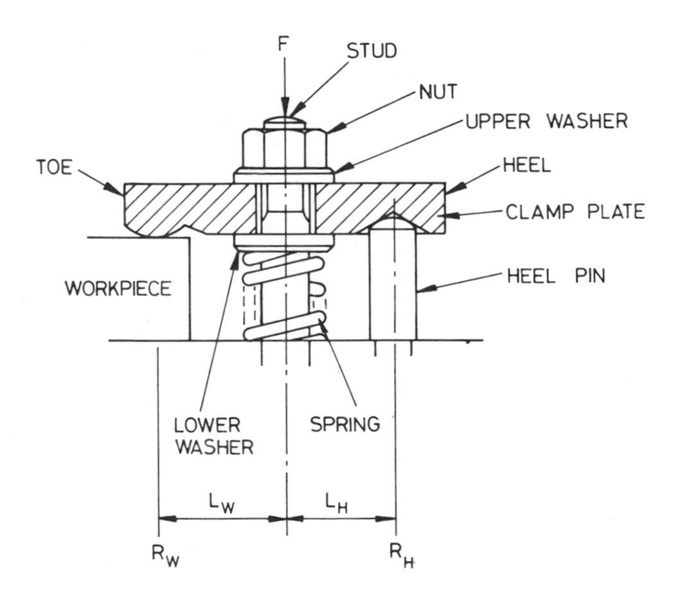

Fig. 1.5 Typical plate clamp

Having decided upon the method of varying the $L_H:L_W$ ratio the system whereby its effect is shown must be considered. As the object of the clamp is to prevent the workpiece from moving, one's first thought may be to devise a system in which a block is clamped and the force required to slide it from under the clamp is measured. Such a system is unsuitable because the clamping force may be so great that it is impossible to move the block and also because the force will no doubt vary from one demonstration to the next so that no conclusions can be drawn. A second, and more effective, scheme is to consider the variations of the reaction at the heel pin (R_H) and at the workpiece (R_W) and to relate them to the $L_H:L_W$ ratio. The reactions can very readily be assessed and the ratio of the two quantities at each stud position is independent of the clamping force.

One method of comparing the two reactions is to use two compression springs of equal stiffness, one at the toe end of the clamp (in place of the workpiece) and the other at the heel end (in place of the heel pin), and to compare the distance that they are compressed at each stud setting. The spring may be protected by enclosing each in a cylinder with a moving cover (see fig. 1.6). By marking the outside of each cylinder with a scale, over which

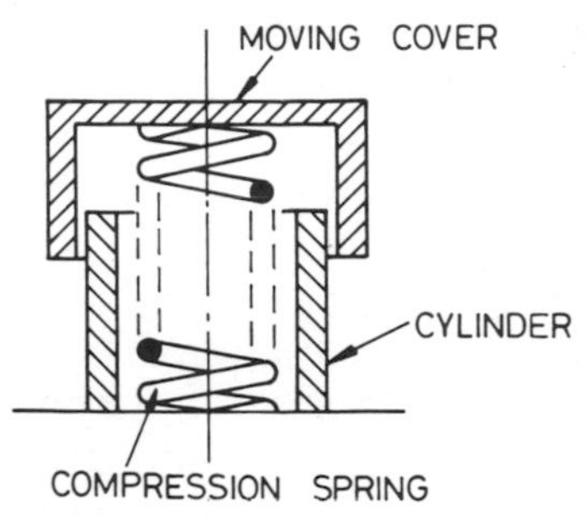

Fig. 1.6 Spring cover

the cover slides (see fig. 1.7), the movement of the two springs, and hence the two reactions, can be compared. The springs must be selected in conjunction with the design of the hand nut so that the movement of the cover is large enough to be seen. A spring of such a small stiffness will require support as shown in fig. 1.8.

As the stud is moved between observations it will be wise to locate the clamp plate with respect to the two cylinders using a system similar to that shown in fig. 1.9.

The complete design scheme, from which the rig can be developed, is shown in fig. 1.10.

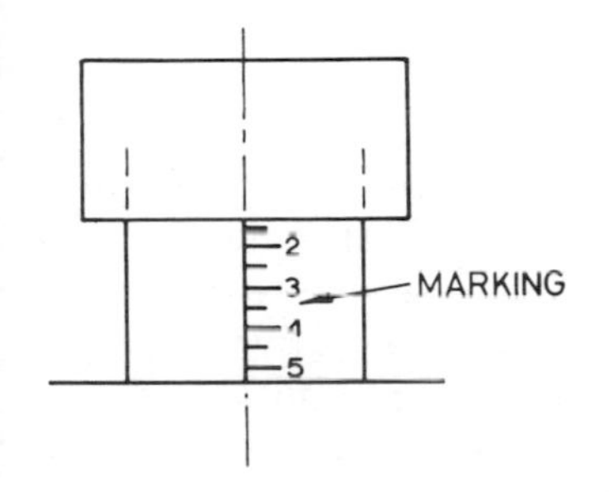

Fig. 1.7 Cover marking

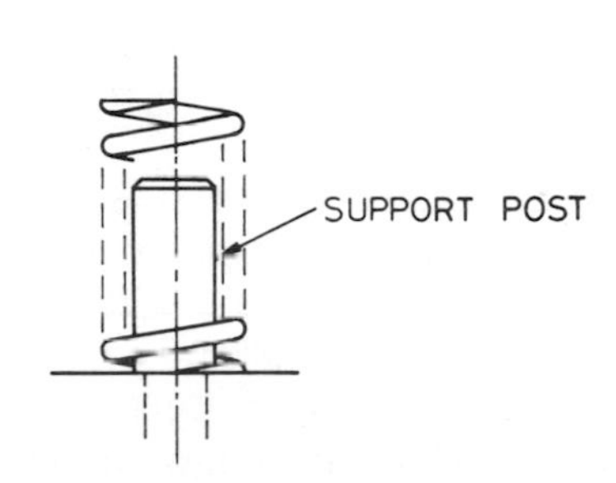

Fig. 1.8 Spring support

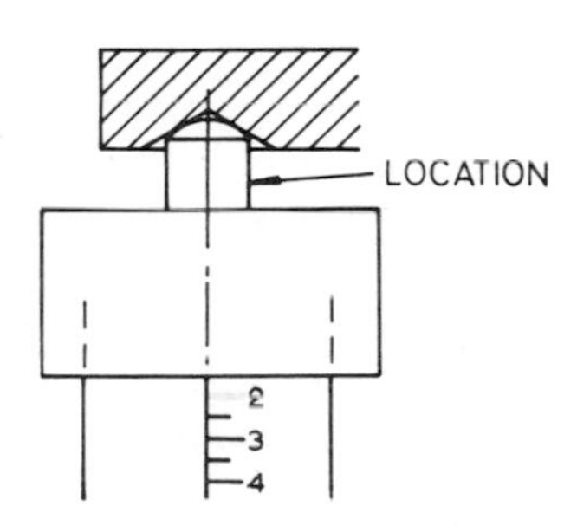

Fig. 1.9 Spring location

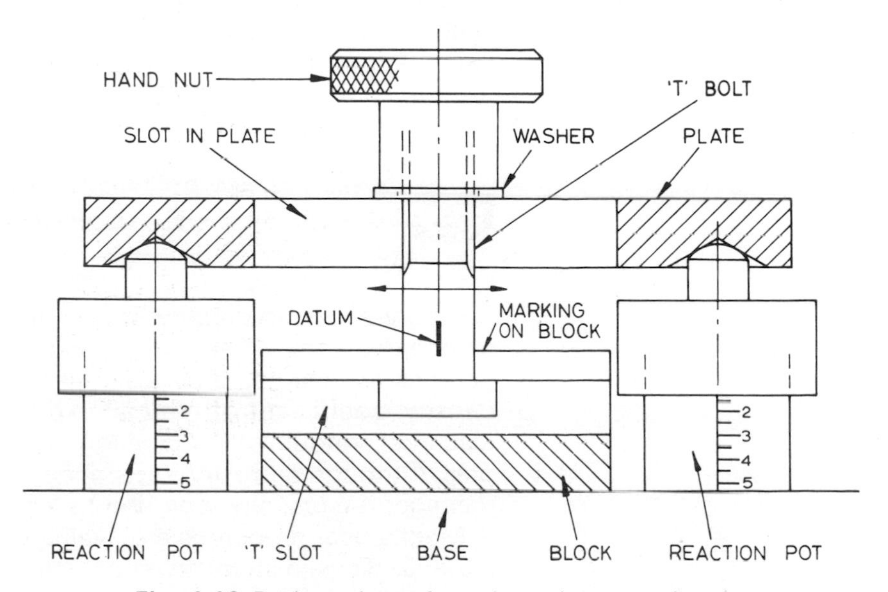

Fig. 1.10 Design scheme for a clamp demonstration rig

2 Selection of Material

The first problem, when selecting the material from which a product is to be made, is to define the requirements that are associated both with its performance when in service and with its manipulation into the shape of the product. The second problem is to obtain a balance between these requirements and the properties and other characteristics of the materials that are available, with due regard to the economic considerations. The shape of the product and the selection of the manufacturing method (see chapter 3) must also be considered when the material is selected. It is assumed that the reader has a sound knowledge of properties of materials and is familiar with engineering materials and their thermal and mechanical treatment. Materials and their properties will be cited, without definitions or explanation, to illustrate the selection of material.

2.1 Physical and mechanical properties

It is convenient to classify properties as physical properties and mechanical properties. Physical properties are those that, in general, are evaluated by tests that do not damage the material: density, coefficient of thermal expansion, thermal conductivity, electrical conductivity and magnetic properties are typical examples. Although corrosion resistance cannot be evaluated without damage, it is usually regarded as a physical property.

Mechanical properties are evaluated by tests that damage the material. Tensile strength, shear strength, compressive strength, hardness, ductility, toughness, resistance to reduction of strength at elevated temperatures, fatigue resistance and creep resistance are typical mechanical properties. British Standard Specifications for materials quote the composition, the accepted impurities and the mechanical properties (usually tensile strength, ductility and hardness) that define the quality of the material. Other mechanical properties, physical properties and characteristics (such as those related to manufacture) are usually the subject of data produced by the manufacturer of the material and the various materials development associations.

2.2 Costs

The price of raw materials is not included in British Standards Specifications because those documents are instruments for the control of quality only and because the price fluctuates with changes in labour and fuel costs, changes in the exchange rate, changes in the international political scene, and with supply and demand. The current prices are given in market reports and in manufacturers' quotations.

Alloys with special properties such as high strength, heat resistance and corrosion resistance are usually expensive because of the high cost of the alloying additions and of the refining and alloying processes. They should only be used when there is no alternative solution. Very often, by careful design, a less costly material can be used. For example, when an increased mass and size is acceptable, strength can be increased by making a component of thicker section to reduce the stress instead of using a material of higher strength. Similarly, an acceptable degree of corrosion resistance may be obtained by surface treatment.

The cost of material when manipulated into castings, forgings etc. depends upon the complexity and accuracy of the product, upon the total quantity to be produced and upon the rate of production as well as the price of the raw material. This cost is often influenced by the supplier's 'order book situation' and by financial agreements relating to the placing of other orders.

When the material from which to make a part is being selected, the total cost of production must be considered. The total cost includes the cost of finishing operations such as machining and joining, and the related inspection work. It may be wise to use a more costly material if the total cost is less.

2.3 Requirements and properties of materials

The requirements of materials are, for convenience, classified as service requirements and manufacturing requirements. These must be studied in conjunction with economic considerations, and the materials selected as a result of compromise.

(a) Service requirements

The service requirements are satisfied by certain mechanical properties (such as strength, hardness, toughness and heat-resistance) and by certain physical properties (such as suitable thermal conductivity, suitable electrical conductivity and corrosion resistance).

Even at this stage, a compromise is usually necessary because even though, in a specific instance, the required properties will be few in number, it is unlikely that they will be found in one material. For example, a material cannot be both very hard and very tough. It is usually necessary to accept a compromise, such as settling for a combination of limited hardness and limited toughness, accepting a low toughness (and designing the product accordingly), or, in the case of suitable steels, by surface-hardening to produce a hard surface on a tough core.

(b) Manufacturing requirements

The manufacturing requirements are related to certain physical properties (such as fluidity and a low melting point to assist in casting) and to certain mechanical properties (such as a fairly low hardness, to permit the use of metal-cutting tools, or high ductility to enable the material to be manipulated by cold-working without fracture). Other characteristics, such as ease of joining by welding or response to heat-treatment to increase the strength, may also be required.

As with service requirements, a compromise is usually necessary because it is unlikely that a particular material will be equally well suited to all the manufacturing processes that are necessary to manipulate it into the required shape.

2.4 Application of the study of requirements to the selection of material

The following examples will illustrate how the service requirements and the manufacturing requirements are balanced, and how the economic considerations influence the final choice.

(a) Metal for the bodywork of a medium priced family motor car

The first step is to determine the function of the bodywork. This can be summarised as: (i) to provide some protection for the passengers in the event of an accident; (ii) to protect the passengers and contents against the atmosphere, the flow of air, fumes from other cars and debris from the road; (iii) to prevent the contents falling from the car; (iv) to give some protection for the car and its contents from the opportunist thief; (v) to produce streamlining; (vi) to be the means of introducing styling as a selling point for the manufacturer and as a status symbol for the customer.

The service requirements are therefore suitable strength and good appearance; corrosion resistance is also important to enable strength and good appearance to be retained. The metal need not be of low density in order to obtain a low mass because the body will be of thin sheet metal to combine maximum passenger space with minimum external dimensions.

The cheapest way to produce a complicated shape in sheet metal is to manufacture the body by presswork. As this is a cold-working operation scaling will not occur during the maufacture of the bodywork and so it will not need to be subjected to descaling operations to ensure a good finish before it is primed and painted.

The principal manufacturing requirements are low strength (so that pressing can be easily done), high ductility (to allow the material to be easily deformed without splitting) and ease of joining by spot-welding (so that the bodywork sections can be rapidly and neatly joined).

A high resistance to corrosion is very desirable, but because an aluminium alloy or a stainless steel would increase the cost excessively a measure of corrosion resistance must be obtained by careful design to eliminate features that promote corrosion. Adequate preparation and finishing, added protection (such as the sealing of joints) and the coating of the underside to minimise damage by stones from the road, can all help prevent corrosion.

The metal usually used is low-carbon steel; it is strong enough to satisfy the basic requirements, can be easily pressed into shape without splitting, can be readily joined by spot-welding and can be painted to obtain the necessary customer appeal. Also it has an acceptable degree of corrosion resistance and is reasonably inexpensive. This is a typical example of a situation in which the mechanical and physical properties that are necessary to meet the service requirements are less important than those that satisfy the manufacturing requirements, and where the costing requirements must be met if the product is to be financially viable.

(b) Metal for the body of a carburettor of a motor car engine

The function of a carburettor body can be summarised as (i) to provide a mixing chamber for the air and petrol; (ii) to provide a housing for the jet or jets; (iii) to provide location and support for the throttle and its control mechanism.

The service requirements of the metal from which the body is made are: sufficient strength to resist distortion; resistance to distortion and loss of strength caused by heat from the engine; local hardness to prevent malfunctioning of the throttle mechanism; resistance to atmospheric corrosion; and resistance to attack by petrol.

The manufacturing requirement is ease of casting because of the complexity of the product. The large quantities warrant the use of die-casting, which will almost completely eliminate machining. Metal to be die-cast must have a low melting point, and if the casting is to be done by the hot-chamber system (which is the more rapid system) a zinc-base alloy is the obvious choice.

A zinc-base alloy would be quite acceptable because it is inexpensive and will satisfy all the service requirements if bushes are introduced during casting. These produce the local wear-resistance necessary for trouble-free operation of the moving parts.

In this situation there is no compromise regarding the requirements, and the only problem is that the total quantity required must be large enough to justify the cost of the maufacture of the dies.

(c) Metal for the turbine blades of a gas turbine engine

The function of the turbine blades is to use the energy of the hot gases to rotate the turbine shaft which, in turn, rotates the compressor shaft and any other output shaft.

The service requirements of the metal are high strength (so that the blades can be thin and strong) and resistance to loss of strength, dry corrosion and creep associated with the high temperature at which the turbine will operate.

The service requirements are only met by special alloys, such as nickel-chromium alloys, which are expensive and awkward to manipulate. In this example, the service requirements must be met if the product is to function correctly for a reasonable length of time, and means must be developed to enable the product to be manufactured. The only cost consideration that can be applied, other than finding the cheapest way to manufacture the blades, is to ensure that the customer will pay the necessarily high price for the 'high technology' product.

2.5 Appraisal of alternative materials

In these examples the choice of material was obvious; but in many cases it may be necessary to select the material that is the best compromise. One method whereby alternative materials can be appraised is to list all the requirements and to rate each of the acceptable materials against each of these requirements. The material with the best overall profile is selected. This method, applied to the appraisal of alternative product designs, is described in chapter 1.

3 Selection of the Manufacturing Method

When selecting the method to be used to manufacture a specific product many factors must be taken into account. These include the basic shape and size of the product, its required strength and the material from which it is to be made, as well as the economics of quantity and cost, the time allowed to manufacture the product and the manufacturing equipment that is available in the company (or is available from a subcontractor). The final choice of manufacturing method usually involves compromise.

3.1 Manufacturing methods and their characteristics

Manufacturing methods can be broadly classified in four groups as follows.

(a) Manufacturing by machining stock materials.
(b) Manufacture by casting, forging or by powder metallurgy, followed by machining as required.
(c) Manufacture from sheet metal by handwork or by presswork.
(d) Manufacture by riveting, welding, brazing etc.

(a) Manufacturing by machining stock materials

Manufacture by machining a billet is suitable when only one part is required or when a batch is required in advance of the main production run. It is usually unsuitable for larger quantities because it is slow and costly. Manufacture by machining bar stock or tube is a suitable method whereby small parts can be produced in any quantity. The manufacture of bolts and unions from hexagonal bar using a capstan lathe is a typical example.

(b) Manufacture by casting, forging or by powder metallurgy, followed by machining as required

Casting, forging and powder metallurgy are alternative primary manufacturing systems. The system adopted will depend upon the specific material from which the part can, or must, be made, the shape and size of the part, its required strength, the manufacturing equipment that is available, and the economics of quantity and cost.

The extent of machining depends upon the primary process that is used, the shape of the product and the required accuracy. For example, a carefully designed die-cast product will require little, if any, machining; but a forging almost always requires machining. The aim is always to eliminate machining, or at least to minimise it, because it is expensive and time consuming. Costly machining may be necessary in, for example, the production of a high-strength forging or the manufacture of a critical component to high accuracy.

(c) Manufacture from sheet metal by handwork or by presswork

Sheet metal construction is used to produce thin-walled, box-like structures such as a motor-car body or a washing-machine cabinet as well as thin parts such as a bracket, a clip or a washer. This method allows a part of thin section to be rapidly produced and it does not involve the extensive machining that is necessary when a thin part is manufactured by other methods. A rigid structure can be obtained by careful three-dimensional contouring, but the shape cannot be as complex as that produced by casting, powder metallurgy or machining. Also the material must be sufficiently ductile to enable it to be cold-worked without splitting.

Handwork is used for small quantity production and for very large parts: presswork is used for large-quantity production and for smaller parts. A large part, such as a motor-car body, can be produced by joining smaller parts by spot-welding.

(d) Manufacture by riveting, welding, brazing etc

Fabrication by joining plate, sheet, extruded sections, machined parts etc. is used when the product is too large to be made in one piece, when it is of a form that does not allow its manufacture by other methods or when the manufacturing schedule does not allow sufficient time for the production of castings, forgings, etc. The joining method used must be suitable for the materials and the product must be designed to allow adequate access to the regions where the joining is to be done.

3.2 The selection of the manufacturing method

Several factors must be taken into consideration when selecting the manufacturing method which best satisfies the requirements. The characteristics of the principal manufacturing methods with respect to the more important factors can be summarised as follows.

(a) Product shape, accuracy and size

Casting can produce fairly accurate shapes because molten metal will take up a complicated shape that it will retain when it is solid. The degree of complexity and accuracy that can be obtained, and the maximum size of casting that can be produced, depend upon the casting system that is used. Sand-casting can produce castings of almost any size which can, if necessary, be hollow. The main limitation on the shape is the removal of the pattern from the mould before the metal is poured in. The accuracy of the product is poor when compared with that obtained by other casting systems. Gravity die-casting is similar to sand-casting but a metal mould (called a *die*) is used, enabling a more accurate casting to be produced. The shape is more limited because the product must be removed from the metal die. Pressure die-casting can produce very accurate castings, but such castings cannot be hollow and must be of a suitable shape so that they can be rapidly removed from the die. Also they must be small enough to suit the capacity of the die-casting machine employed. Investment-casting can be used to produce complicated castings that can be hollow if necessary, but these must be of small or medium size.

Powder metallurgy can produce small, accurate parts. The shape of parts produced by this method is limited by the inability of metal to flow easily when in powder form and the need to rapidly eject the compacted powder from the die.

Forging can be used to produce small parts and medium-size parts, but they must be of simple shape because solid metal does not readily flow; hollow parts cannot be produced by forging. The limitations of shape, combined with the inaccuracy associated with hot-working, usually result in the need to remove a considerable quantity of metal by machining to produce an acceptable shape and accuracy.

Sheet-metal construction, or fabrication using sheet metal, machined parts etc., can be used for large products or for those of a shape unsuitable for manufacture by other methods.

Machining can produce complicated and accurate shapes and can finish parts produced by forging, casting or powder metallurgy. Parts can also be machined from billets, bar stock etc. The size of product that can be machined is usually limited only by the capacity of the machine tool.

(b) The required strength of the product

Forging develops a high and directional strength and is the best method to use when strength is important. The strength of parts produced by casting is not high, and that produced by powder metallurgy is low, but by careful product design and selection of material a part with adequate strength can often be produced by casting or by powder metallurgy.

(c) The material from which the product is made

Materials are usually suitable for either casting or working (groups of alloys, aluminium alloys, for example, are classified in this way) and it is essential that the material selected and the manufacturing process to be used are compatible. Many manufacturing processes also require materials with special properties. For example, metal to be die-cast must have a low melting point because metal dies are used, and metal to be manipulated by presswork must be ductile.

(d) Quantity required

Some manufacturing processes, such as pressure die-casting and powder metallurgy, require expensive tooling equipment and can only be profitably employed when a large quantity is required.

(e) Cost of manufacture

The aim is always to obtain the required quality at minimum cost. Sometimes the conditions that will be met when the product is in service, or special requirements of shape, demand that an expensive manufacturing method, such as investment-casting, is used to manufacture the product. Very often the overall cost of using an expensive manufacturing method can be reduced considerably by using its ability to produce a complicated shape. Two or more components can be combined so that a single, but more complicated, component is produced instead.

3.3 Appraisal of alternative manufacturing methods

Certain methods will be eliminated when manufacture is considered. For example a bridge could not be made as a one-piece casting or a cylinder block by forging. Certain methods will, however, be accepted immediately. For example, when hexagon-headed bolts are required they would, without question, be produced by turning hexagonal bar. Sometimes the design is modified to enable a non-traditional manufacturing method to be used. For example door bolts and handles can be made from extruded sections that are cut to length.

There are many shapes that could be made using a number of methods which must be compared before the final choice is made. The following are typical examples.

A crankshaft could be made using a number of methods. One extreme case is the crankshaft for a petrol engine to power a model aircraft; this crankshaft could be made by turning a piece of bar material on a lathe. Another extreme case is the crankshaft for a large engine which could be made up of sections that are bolted together. The crankshaft for a motor car could be made either from a forging or from a casting and it is necessary to compare these two methods before a decision is made. The forged crankshaft would be stronger than the cast crankshaft: but the total cost of the forged crankshaft would be higher than that of a cast crankshaft because the equipment is costly, the forging dies expensive to make, and because a considerable amount of metal must be removed at the machining stage. The weaker cast crankshaft would be less costly because the equipment and the pattern is cheaper, and less material must be removed at the machining stage. The final choice must be made by balancing the strength against the cost.

A complicated casing would most likely be made by casting, and in this case it is the type of casting method that must be considered. Assuming that the casting is fairly small it could be made by sand-casting, by die-casting or by investment-casting. If the shape of the casting is not particularly awkward there is no point in using investment-casting, and so the choice is between sand-casting and die-casting. Sand-casting is suitable for small or large quantities in any casting alloy; die-casting is more suitable for large quantities (because the high cost of the dies makes small quantities uneconomical) and is limited to alloys of lower melting point. Sand-casting does not produce very accurate castings compared with die-casting, which produces castings that require little, if any, machining. The final choice must be made with reference to the material, the quantity required and the rate of production, and the required accuracy and appearance of the casing.

When small components are required, powder metallurgy may be a suitable alternative to die-casting because it can be applied to high melting point materials as well as to low melting point materials (aluminium is the only material that is unsuitable). The accuracy of the product is very high and the rate of production and economics of quantities are similar to those associated with die-casting, but for a specific material and shape the product of powder metallurgy is less strong than that of other manipulation methods.

There are occasions when a component can be produced by welding together pieces of plate, tube, angle section etc. as an alternative to casting. For example, a set of four brackets is urgently required to replace an existing set. The brackets are bolted to an outside wall where they support a shaft that runs parallel to the wall and the shaft is slowly turned by hand. The brackets do not therefore need to be particularly strong, they can be reasonably heavy, and their appearance does not matter. They can be produced as iron castings or by welding together pieces of mild steel plate and a length of mild steel tube (the latter to form the bearing portion). In both cases machining will be necessary to obtain the required accuracy. In view of the urgency and the small quantity that is required the fabricated product is the better one because, by careful design, it can be made from readily available material and would possibly be completed even before the pattern for the casting. The cost of the fabricated product would be lower than that of a cast bracket and, as stated, the less attractive appearance is acceptable.

When the choice of method is not immediately obvious a rating system that is similar to that described in chapter 1 may be used.

3.4 Detail design to suit the manufacturing method

When the manufacturing method has been selected, the product must be designed to conform with the characteristics of the method and to fully exploit its merits. Detail design to suit the basic manufacturing methods is the subject of the chapters that follow but it must be emphasised that the manufacturing method should be decided at the initial design stage so that the manufacturing requirements can be considered at all stages in the design of the product. They must not be left to the detail design stage when it may be too late, or be too costly, to introduce modifications to minimise the manufacturing problems.

4 Design for Casting

The factors that influence the selection of the manufacturing method are discussed in chapter 3. In this chapter the problem areas associated with the principal casting systems are identified and the detail design of castings is related to them. It must be emphasised that casting design should be considered at the design scheme stage, where very often a simple modification to simplify casting can be easily accommodated. Such modifications might be troublesome to accommodate at the detail design stage if adjacent components were affected.

4.1 Design of sand-castings

The problem areas associated with sand-casting can be identified as:

(a) the manufacture of the mould;

(b) the behaviour of the metal during pouring, solidification and cooling in the solid state;

(c) the fettling of the casting.

In addition to these problems, the cost of producing and machining the casting must be considered. Features that increase the cost of moulding and fettling and those that demand additional machining operations should, if possible, be avoided.

Hollow castings can be produced by including a sand core to restrict the space into which the metal can flow, but if the core is a separate piece the cost of moulding will increase. Therefore although a characteristic of sand-casting is its ability to produce hollow, box-like shapes, a less complicated shape should be used if possible.

Very often one design change will affect more than one

problem area; but in the following sections the problem areas are considered individually.

(a) Design to simplify moulding

The reader should refer, if necessary, to a suitable text book dealing with sand-casting. Basically the moulding operation consists of packing sand around the pattern (which resembles the un-machined casting), removing the pattern from the mould and assembling the sections of the mould. The mould is usually in two parts and the pattern, or part-pattern, is seated on a board or is mounted on a plate when each part of the mould is produced.

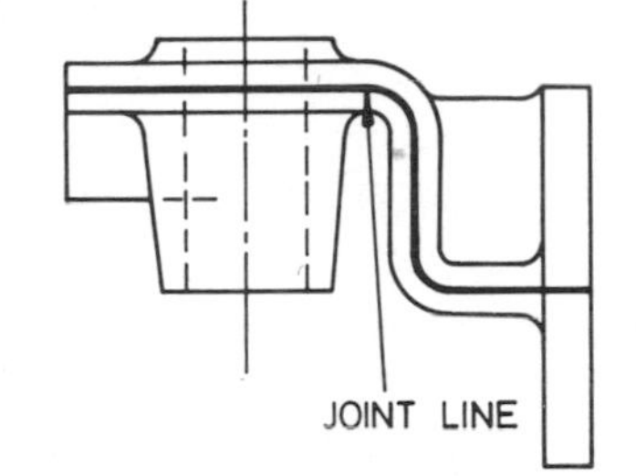

Fig. 4.1 Casting that requires a cranked joint line

(i) The location and shape of the joint surface. The joint surface between sections of the mould must be such that the pattern can be removed after the moulding has been done. As this will form the surface upon which the pattern, or part-pattern, will sit at the start of moulding it should, if possible, be a plane surface. When designing a casting the joint line should be fixed at the start.

Fig. 4.1 shows a casting that can only be produced by splitting the mould in the direction shown but which, because of the position of the circular flange, requires a cranked joint line. When the flange is re-positioned, as shown in fig. 4.2, the joint line is straight and the pattern can be seated on a plane surface, simplifying the moulding operation and enabling a higher degree of accuracy to be obtained.

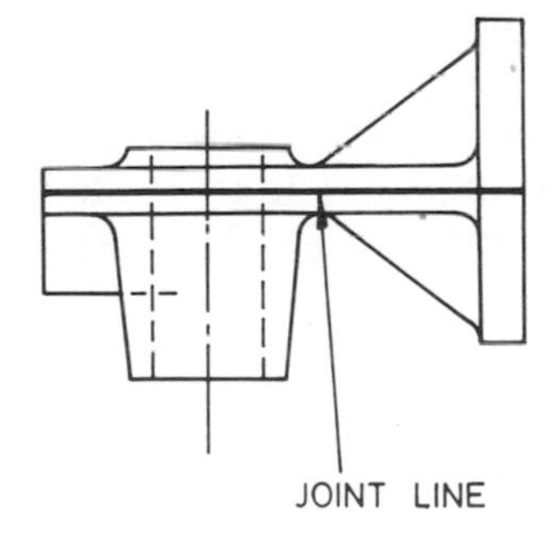

Fig. 4.2 Casting that requires a straight joint line

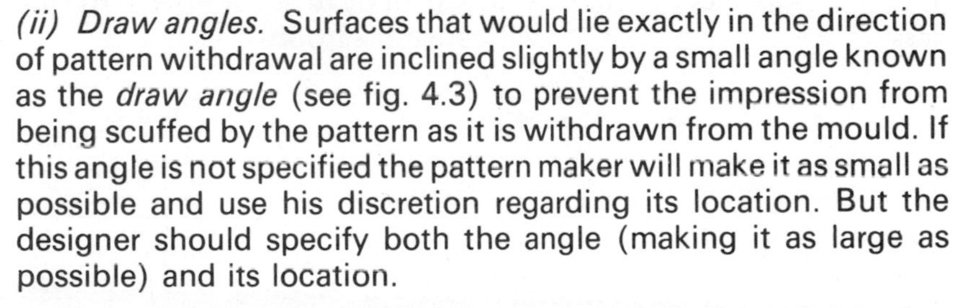

(ii) Draw angles. Surfaces that would lie exactly in the direction of pattern withdrawal are inclined slightly by a small angle known as the *draw angle* (see fig. 4.3) to prevent the impression from being scuffed by the pattern as it is withdrawn from the mould. If this angle is not specified the pattern maker will make it as small as possible and use his discretion regarding its location. But the designer should specify both the angle (making it as large as possible) and its location.

(iii) Re-entrant shapes. Re-entrant shapes should, if possible, be avoided because they require special arrangements to enable

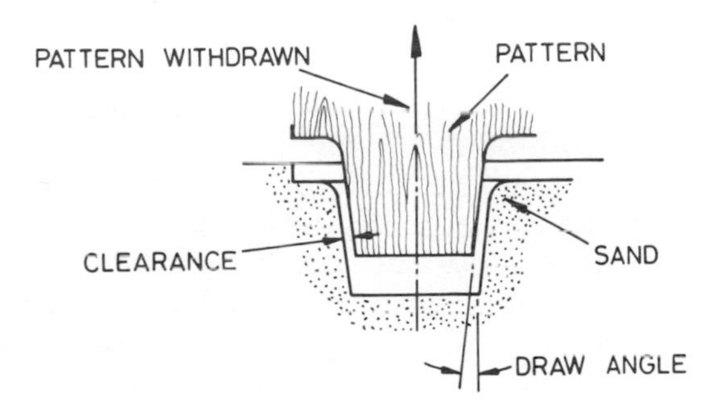

Fig. 4.3 The function of the draw angle

the pattern to be removed. This increases the cost of the moulding and may reduce the accuracy of the casting.

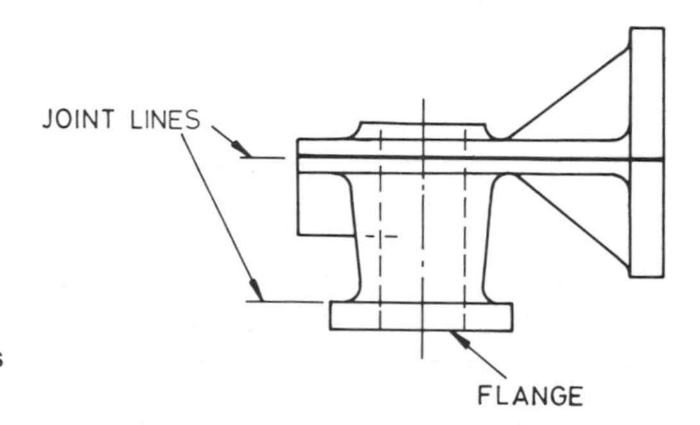

Fig. 4.4 Casting that requires a three-part mould

For example, if the casting under consideration included the flange as shown in fig. 4.4, the pattern and mould would need to be split in two places, requiring a three-part mould, so that the pattern could be removed.

Similarly, the projecting side-boss shown in fig. 4.5 would require either a loose piece (part of the pattern that remains in the mould after the main body of the pattern is withdrawn, to be removed separately) or a core that is positioned in the impression produced by a simplified pattern that does not include a re-entrant feature. This problem is eliminated either by making the boss elongated, as shown in fig. 4.6, so that the pattern is easily removed, or by re-locating it at the joint line (see fig. 4.7) where it will cause no trouble.

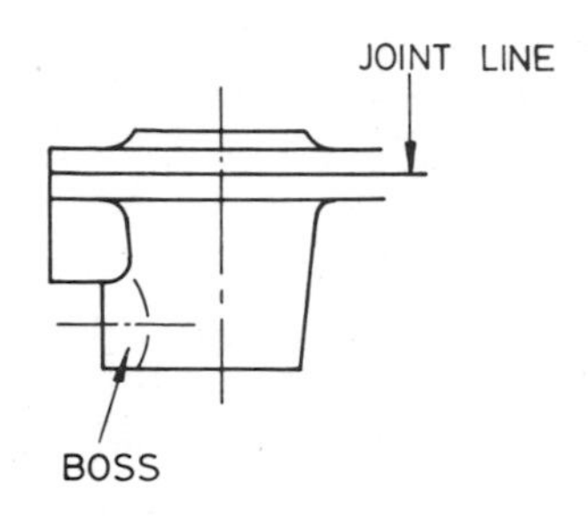

Fig. 4.5 Casting with a side boss

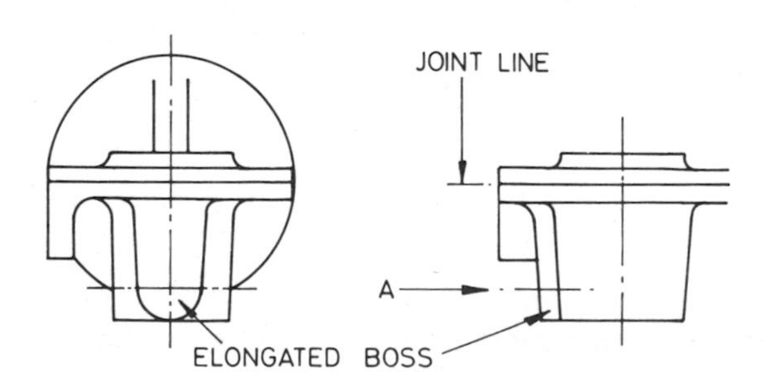

Fig. 4.6 Casting with a modified side boss

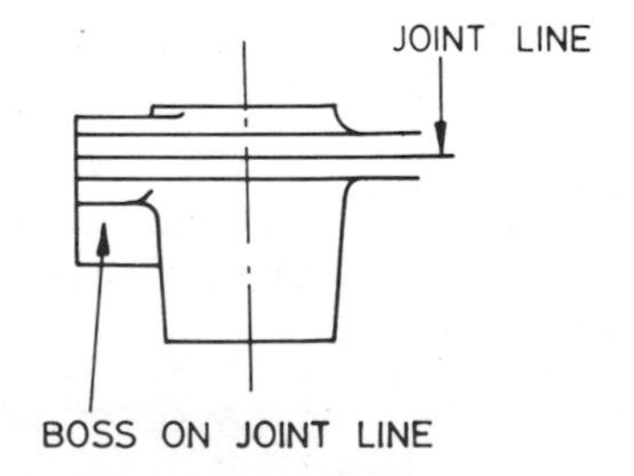

Fig. 4.7 Casting with a boss at the joint line

(iv) Location of bosses. Bosses are often included to cut down the amount of machining that is required, by reducing the machined surface area, as an alternative to localising by spotfacing. Every boss that lies on an upper, horizontal face will require a riser to allow air to escape and so ensure that a complete casting is produced. These additional risers will increase the moulding time and also the fettling time because the metal in them will need to be removed; there will, therefore, be a break-even point at which

the introduction of more bosses becomes uneconomical. Fig. 4.8 illustrates part of a casting that includes such a large number of bosses that it becomes more effective to combine them to form a single facing as illustrated in fig. 4.9.

(b) Design to control the behaviour of the metal during the sand-casting process

In order to appreciate the relationship between casting design and the behaviour of the metal, the pouring and the solidification of metal will be briefly studied at this point.

To ensure continuity of structure and completeness of the product the molten metal must reach all parts of the impression before solidification starts. The contraction of the molten metal during cooling will, provided that the casting design and the mould construction is correct, be compensated by the column of molten metal in the runner(s) and riser(s) acting as feeders.

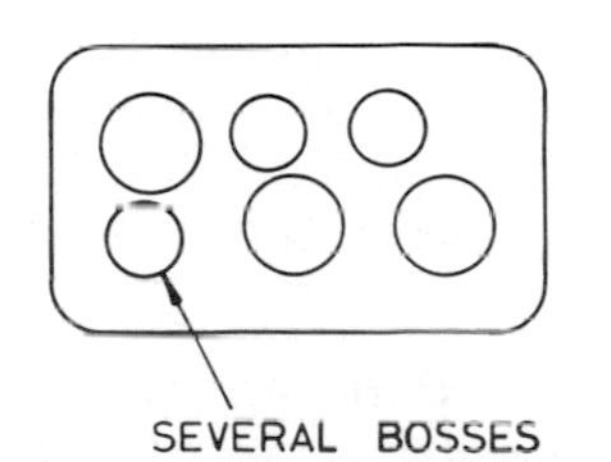

Fig. 4.8 Casting with a large number of bosses

When solid, metals usually consist of grains and grain boundaries: the former are orderly three-dimensional arrays of atoms and the latter are hapazard jumbles of atoms, about three atoms thick, joining the grains. The grains and grain boundaries of an alloy consist of a mixture or a compound of the metals (or metals and non-metals) of which the alloy is composed. The grain boundaries are, when at room temperature, stronger than the grains, and it can be shown that a structure that consists of small grains (a *fine structure*) is therefore stronger than one that consists of large grains (a *coarse structure*). A fine structure is produced by rapid solidification and a coarse structure by slow solidification. The solidification rate is controlled by (i) the mould material, thickness and temperature, and (ii) the thickness of the casting being produced.

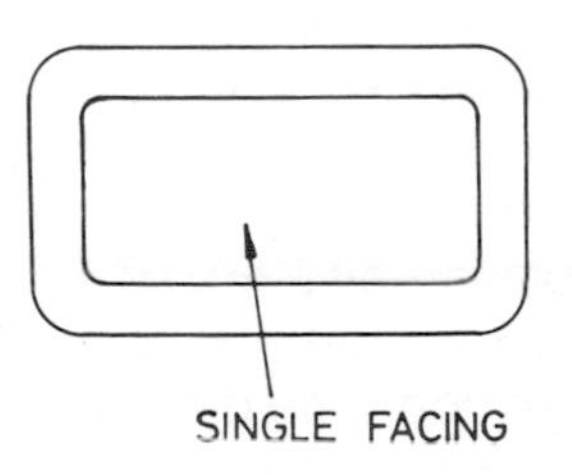

Fig. 4.9 Casting with a single facing

The phases present at room temperature may also depend upon the solidification rate. For example grey cast iron (cast iron that contains graphite) is produced by slow solidification and white cast iron (cast iron that does not contain graphite) by rapid solidification.

As the grain size and, in some instances, the phases present depend upon the casting thickness the casting should, if possible, be designed so that it is of uniform wall thickness if uniformity of structure is required.

The structure also varies across the section because at the outside, where solidification starts and cooling is rapid, the grains will be fine (the grains here are called *chill crystals*) and at the centre of a thick casting, where the cooling will be slow, they will be coarse. Just below the surface, where the grain formation during solidification is more restricted in directions tangential to the surface than normal to it, the grains will be columnar in shape.

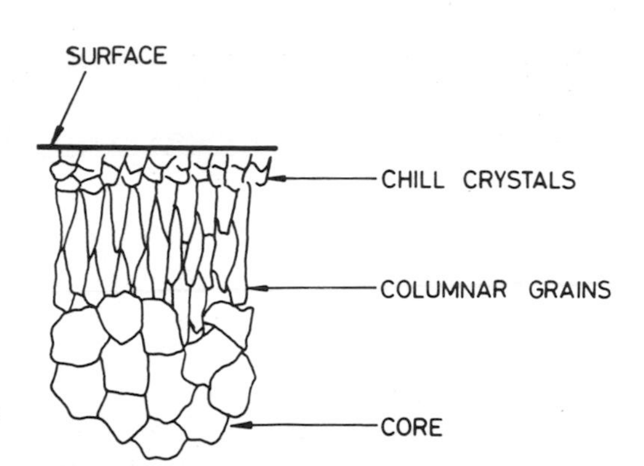

Fig. 4.10 Types of grain in a thick casting

Fig. 4.10 illustrates the three main types of grain that are present in a thick casting.

Contraction during cooling in the solid state is allowed for by making the pattern oversized to produce a larger impression, but when the casting has walls that are not the same thickness the thin parts will solidify first, and they may crack as a result of the stresses imposed upon them when the thicker sections are solidifying.

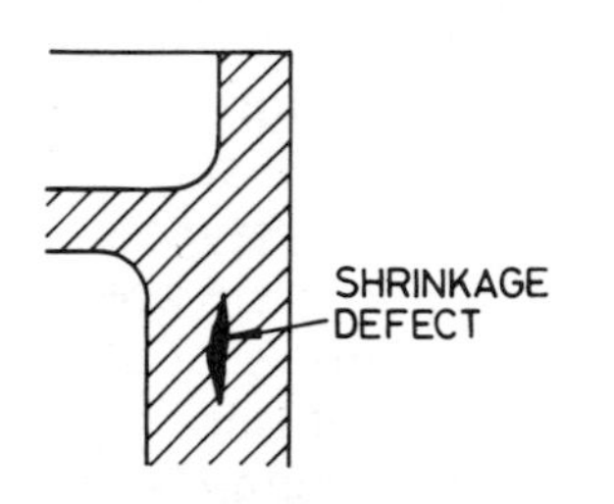

Fig. 4.11 Casting design that produces shrinkage defects

(i) Section thickness. It has already been stated that contraction of the molten metal during the solidification will be allowed for by supply from the runners/risers, and it will be appreciated that a thick section requires more molten metal to allow for contraction than does a thin section. This causes problems when it is necessary to feed a thick section through a thin section; the metal in the thin section will solidify before that in the thick section, cutting off the supply of molten metal and producing the shrinkage defect shown in fig. 4.11. Fig. 4.12 shows how the casting can be re-designed to prevent the defect from occurring.

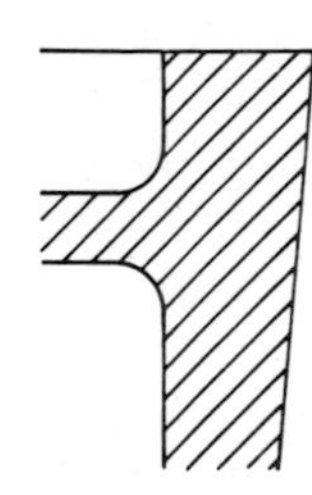

Fig. 4.12 Casting design to prevent shrinkage defects

Instead of making a casting of thick section to obtain greater strength and rigidity, it may often be made of larger 'outside' shape and hollow to reduce the section thickness. Similarly the thick section shown in fig. 4.13 can be replaced by a rib, as in fig. 4.14, to obtain local support. The tall bosses on the flange in fig. 4.15 can be modified as shown in fig. 4.16 to combine height with uniform section.

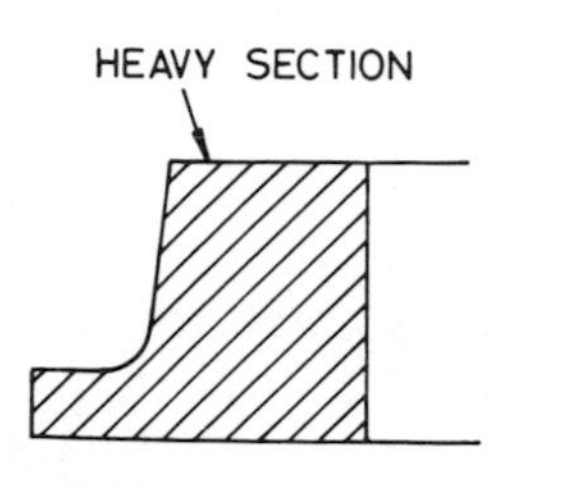

Fig. 4.13 Casting with heavy section

(ii) The junction of wall sections and of ribs with wall sections. The effect of section thickness, and also the variation of grain shape shown in fig. 4.10, must be considered when the casting includes junctions.

Fig. 4.17 illustrates how a plane of weakness is produced when two sections join at right angles, and fig. 4.18 illustrates how this can be eliminated by making the surfaces curved at the intersection. In addition to the weakness associated with the grain shape, the sharp angle at the inside of the casting (fig. 4.17) causes that part of the casting to solidify more slowly because the heat cannot

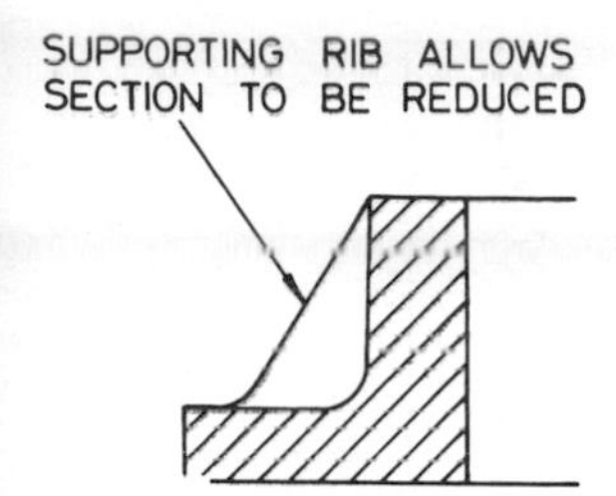

Fig. 4.14 Casting with supporting rib

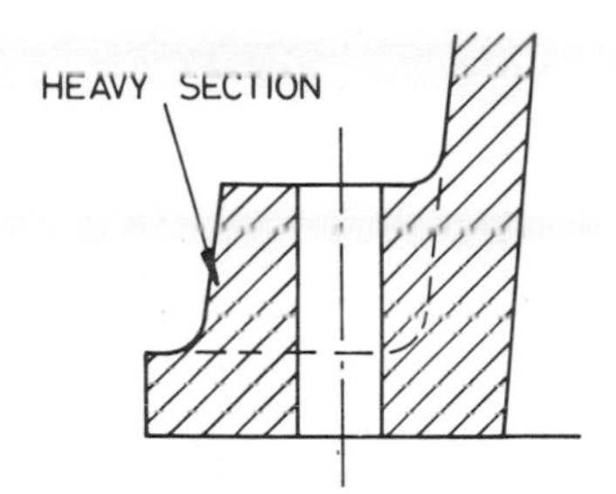

Fig. 4.15 Tall boss causing heavy section

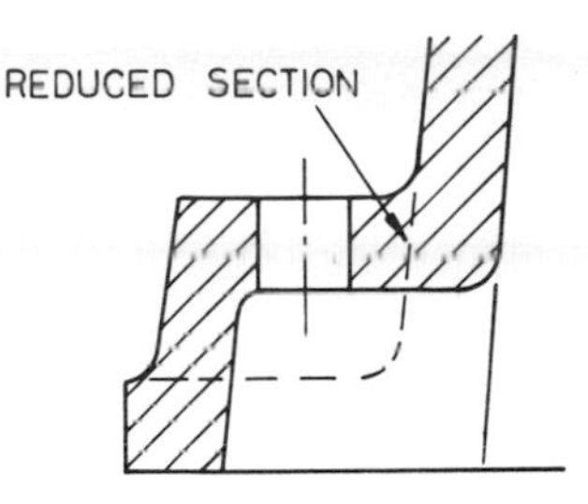

Fig. 4.16 Modified casting

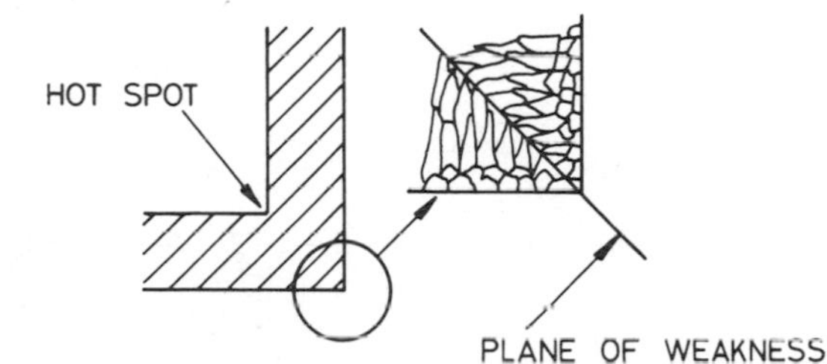

Fig. 4.17 Design causing a plane of weakness and a hot spot

escape so easily into the core, producing a *hot spot*. This is reduced considerably in the design in fig. 4.18, but is further reduced by making the section locally thinner, as shown in fig. 4.19.

When it is necessary to design a casting to have different thicknesses of wall section, there should be a gradual change as illustrated in fig. 4.20.

The hot-spot effect is made worse when sections and ribs join at an acute angle, as in fig. 4.21; the acute angle junction can be retained if the shape is 'opened out' (see fig. 4.22) to allow the heat to escape more readily into the core of the mould.

The extent of the variation of metal mass produced at junctions can be indicated by the *inscribed circle system* (see fig. 4.23) in

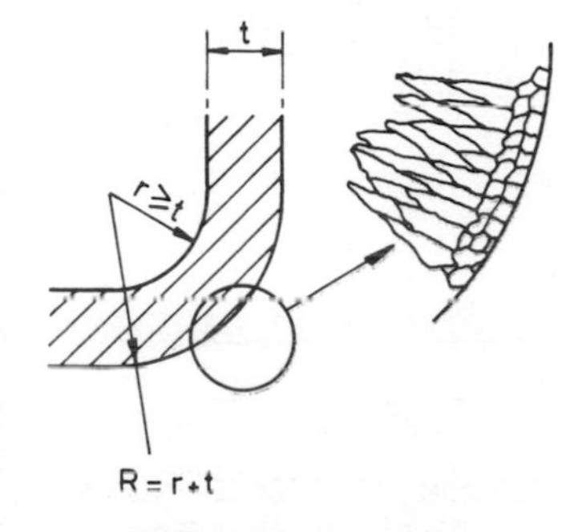

Fig. 4.18 Design to eliminate the plane of weakness and hot spot

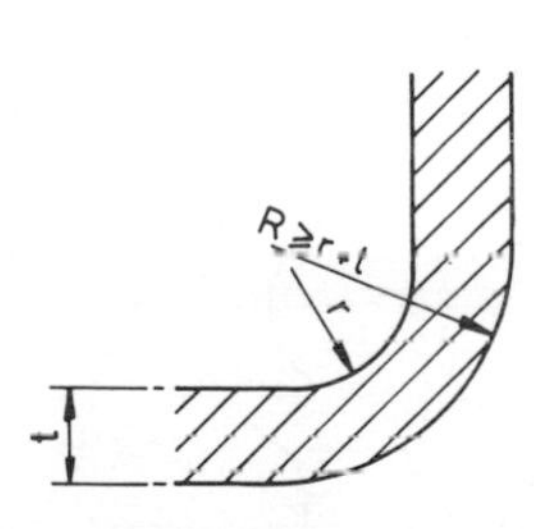

Fig. 4.19 Design to further reduce the hot spot

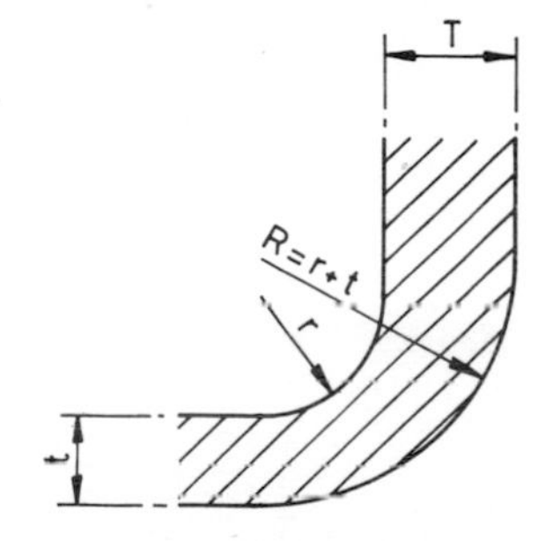

Fig. 4.20 Design to blend different thickness sections

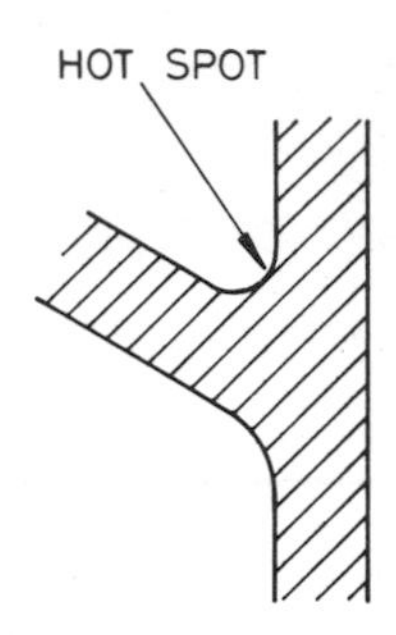

Fig. 4.21 Design that causes a hot spot

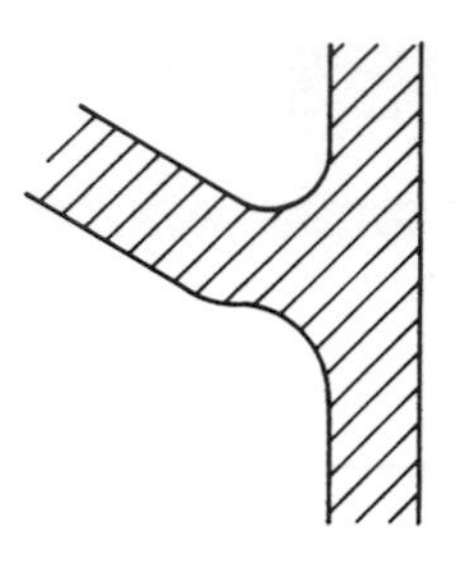

Fig. 4.22 Design to minimise the hot spot

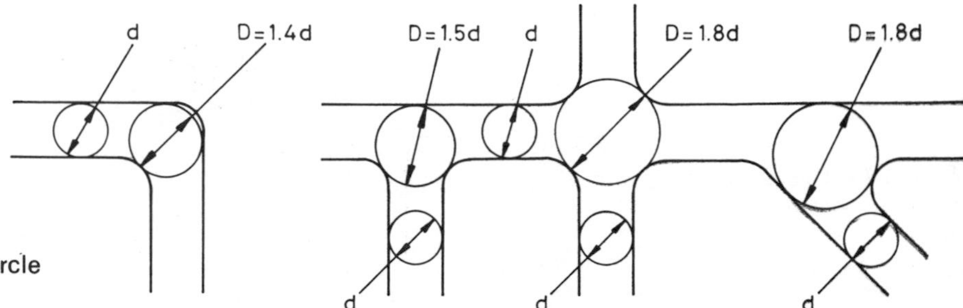

Fig. 4.23 The inscribed circle system

which the mass at the junction (represented by D) is compared with that of the sections that join (represented by d). It will be seen that the variation increases with the number of sections that meet and as the angle of the junction becomes more acute. The mass at a junction can be reduced by using a *cored boss*, as shown in fig. 4.24, but the introduction of the core increases the cost of the moulding operation. An alternative is to reduce the thickness using a local depression (see fig. 4.25) and to combine this with *staggering* when several ribs are involved (fig. 4.26). The local depressions may take the form of corrugations. For example, the mass effect of several ribs supporting a central extended boss (in fig. 4.27) can be reduced as shown in fig. 4.28.

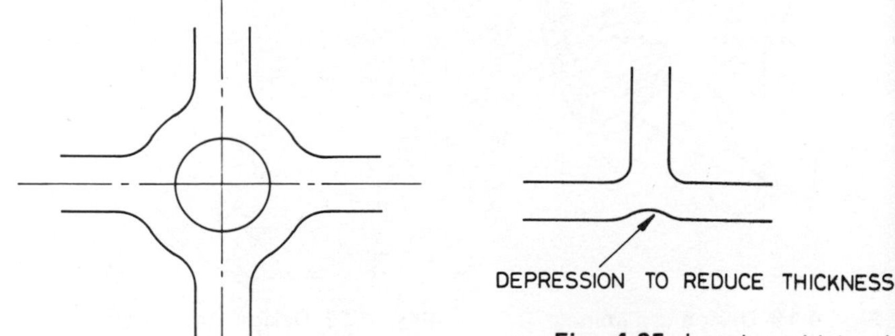

Fig. 4.24 Cored boss

Fig. 4.25 Junction with local depression

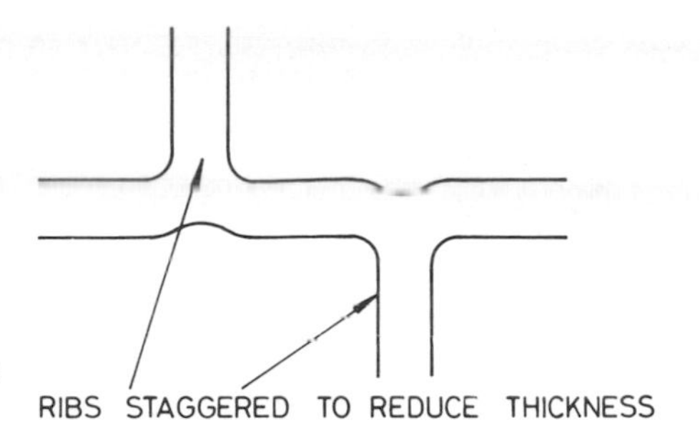

Fig. 4.26 Local depressions and staggered ribs

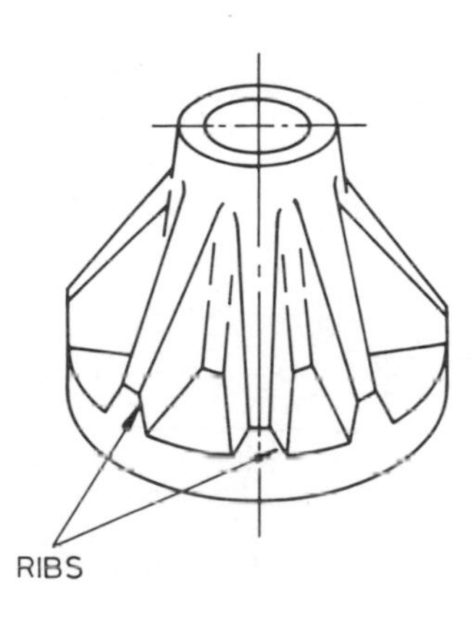

Fig. 4.27 Design that causes mass effect

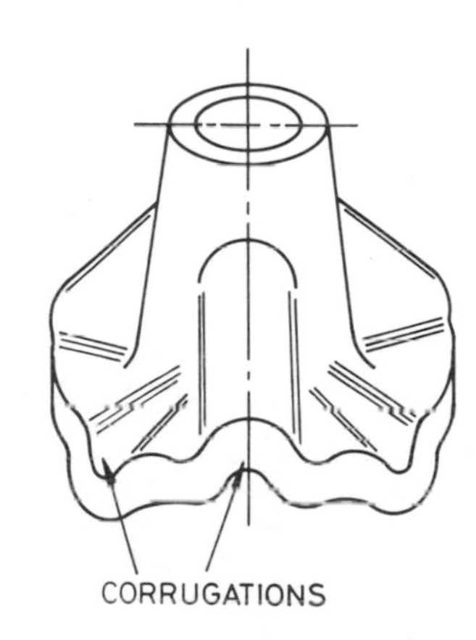

Fig. 4.28 Modified design

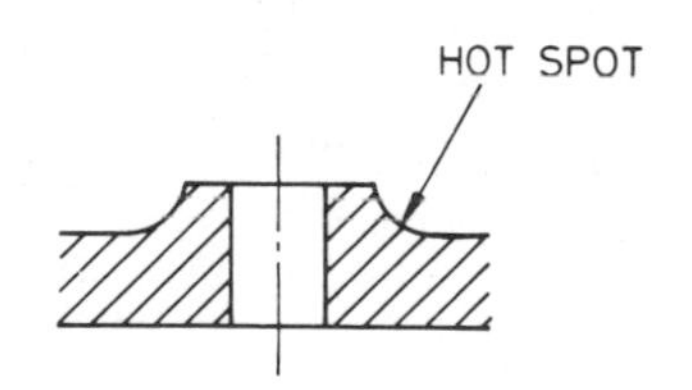

Fig. 4.29 Boss causing a hot spot

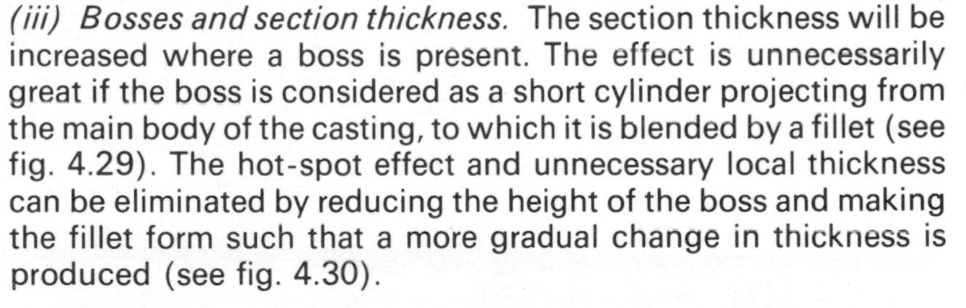

(iii) Bosses and section thickness. The section thickness will be increased where a boss is present. The effect is unnecessarily great if the boss is considered as a short cylinder projecting from the main body of the casting, to which it is blended by a fillet (see fig. 4.29). The hot-spot effect and unnecessary local thickness can be eliminated by reducing the height of the boss and making the fillet form such that a more gradual change in thickness is produced (see fig. 4.30).

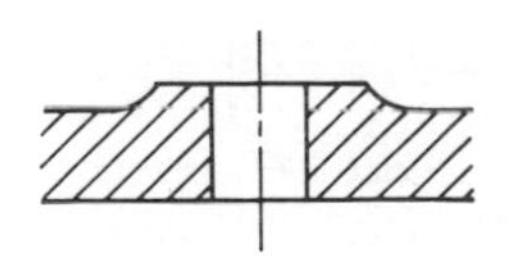

Fig. 4.30 Modified boss

(iv) Separate cores. When separate cores have to be used, gas from the molten metal should be allowed to escape through them in the same way as it is allowed to escape through the mould; this can only occur if the cores extend into, or beyond, the mould. When a core extends in this way it can be located when the mould is assembled and supported in position during the pouring stage. It will produce a *cored hole* in the wall of the casting through which the core sand can be extracted during the fettling. If a cored hole cannot be permitted in the finished casting it may be filled by a *core plug*. This is a threaded part that is screwed into the cored hole (which is threaded for the purpose), locked in position by a dowel and finally machined to remove the slot, square, etc. that

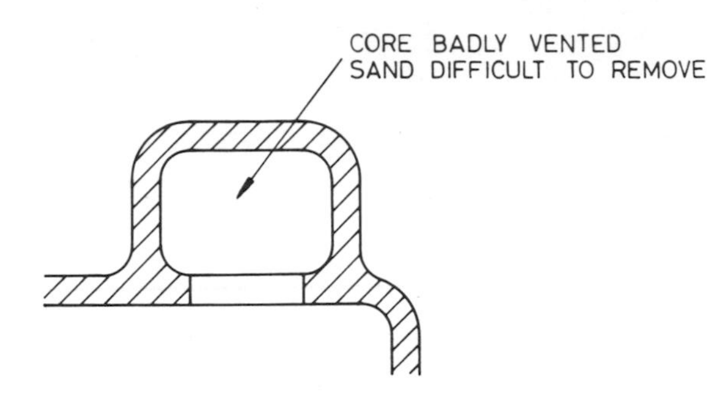

Fig. 4.31 Badly-designed core

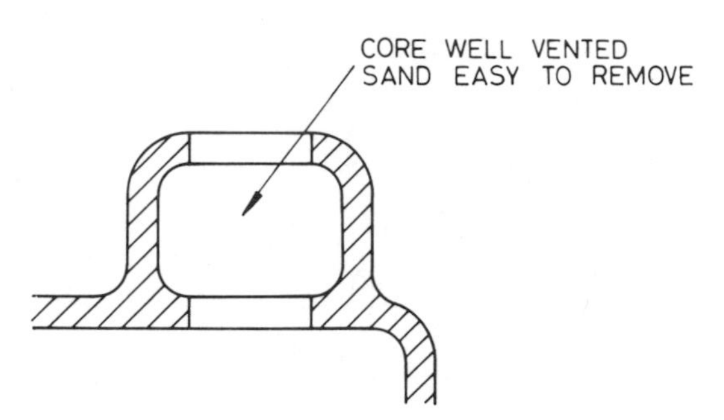

Fig. 4.32 Correctly-designed core

was used to screw it into position (and which could be used to remove it).

Fig. 4.31 shows a design in which the core is badly vented and the core sand awkward to remove. Fig. 4.32 shows a variation in the design eliminating these problems. Care should also be taken to ensure that the core does not move or break under its own weight. Fig. 4.33 illustrates part of a casting that requires a vertical core and a projecting horizontal core; by changing the design to that shown in fig. 4.34 the problems associated with the horizontal core are eliminated.

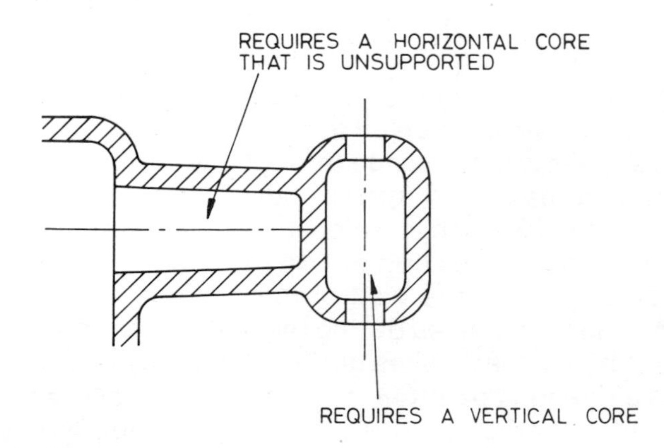

Fig. 4.33 Casting with an unsupported core

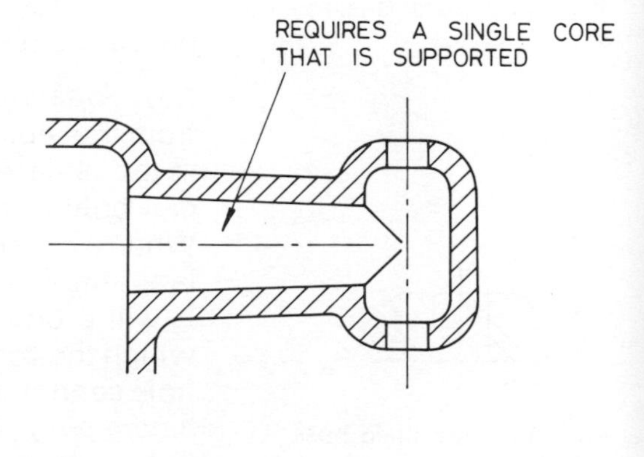

Fig. 4.34 Casting with a supported core

(v) Prevention of stresses and possible fracture during cooling in the solid state. Contraction of the solid metal as it cools to room temperature cannot be avoided and so the casting must be designed with this in mind. For example, when a casting is ribbed to combine stiffness with minimum weight there is a tendency for the ribs to crack as a result of contraction when they are arranged as in fig. 4.35. This cracking can be prevented by arranging them as in fig. 4.36.

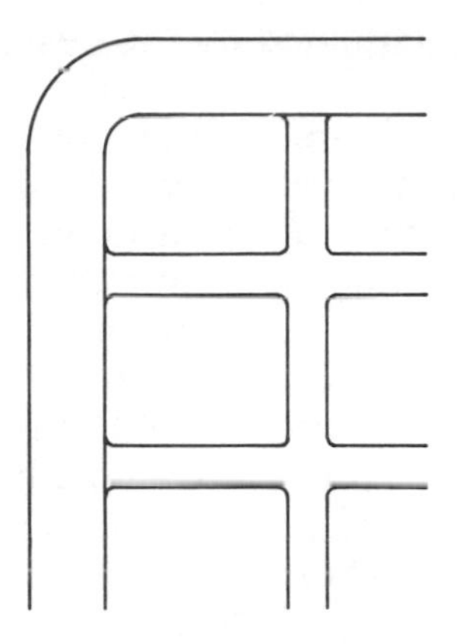

Fig. 4.35 Ribs that may crack during cooling

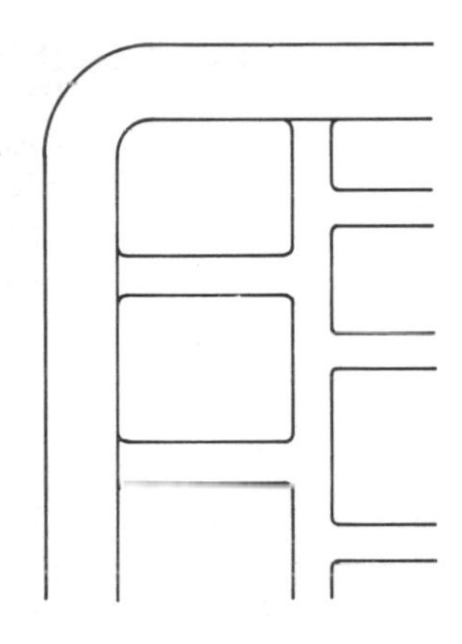

Fig. 4.36 Rib arrangement that will absorb contraction

This can be extended to include the spokes of a ring (for example a cast, spoked flywheel). When there is an even number of straight spokes, as shown in fig. 4.37, distortion during the contraction will cause cracking; but when there is an odd number of curved spokes, as in fig. 4.38, each one can take up a greater or a smaller curvature, depending upon the distortion, without itself cracking or causing cracking to occur elsewhere.

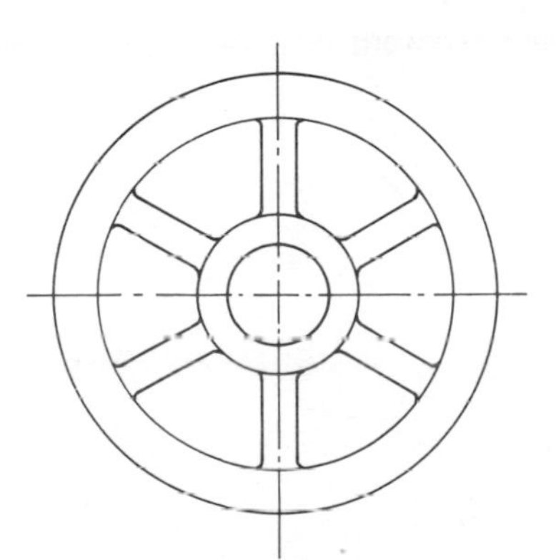

Fig. 4.37 Spokes that may crack during cooling

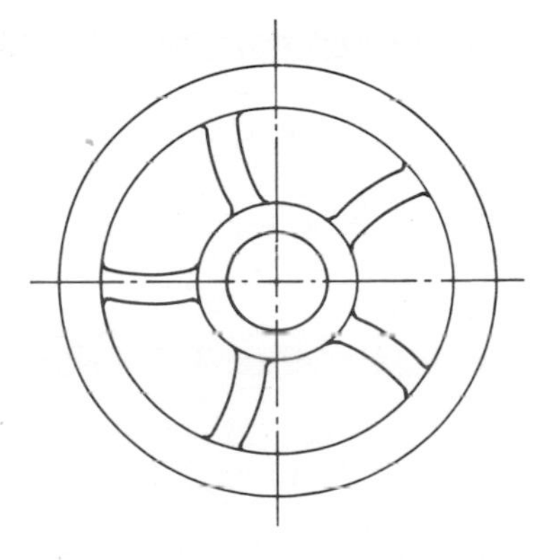

Fig. 4.38 Spoke arrangement that will absorb contraction

(c) Design to assist fettling

Fettling consists of removing the runner(s) and riser(s) and the core sand from the cavities produced by them and generally preparing the casting for the customer, machine shop etc.

Many of the design improvements discussed in previous sections affect the fettling operations but the following deserve further comment.

(i) Location of runners and risers. The location of runners and risers may be determined by the design of the casting because it may be necessary to provide a riser to allow air to escape from the mould and so ensure that a complete casting is produced. Similarly a runner or riser may be necessary at a heavy section to allow for the contraction of the molten metal. The metal that is left in the riser and runner channels (this metal is also called runners and risers) must be removed from the main body of the casting by a blow or by cutting. Every riser or runner will increase the fettling time and so their number should be minimised. It is essential that the runner or riser be removed without damaging parts of the casting in the vicinity. Fig. 4.39 shows part of a casting that includes a boss that lies in a recess, making the removal of the riser very awkward. The operation can be simplified by reducing the depth of the recess, as illustrated by fig. 4.40, or, alternatively, by increasing the height of the boss.

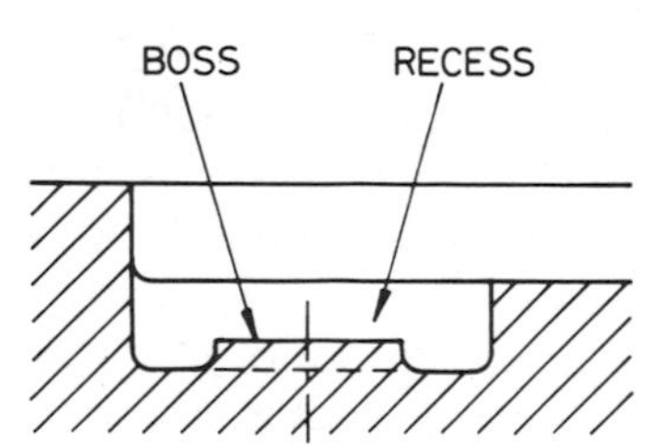

Fig. 4.39 Design that will make fettling awkward

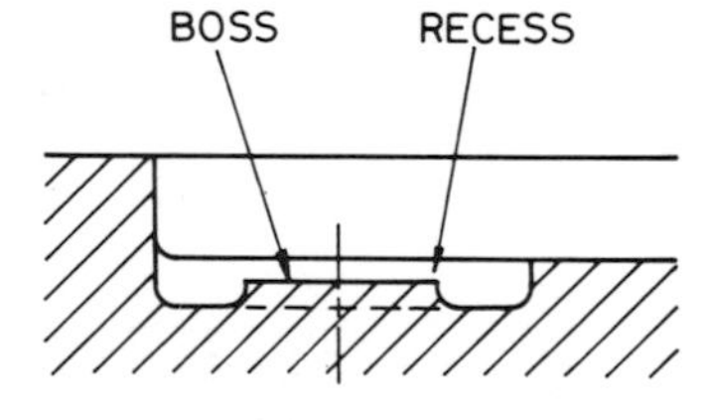

Fig. 4.40 Design that will simplify fettling

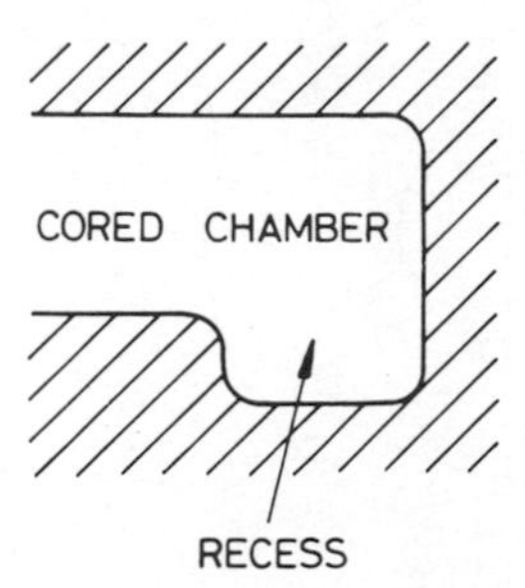

Fig. 4.41 Design in which sand-removal will be difficult

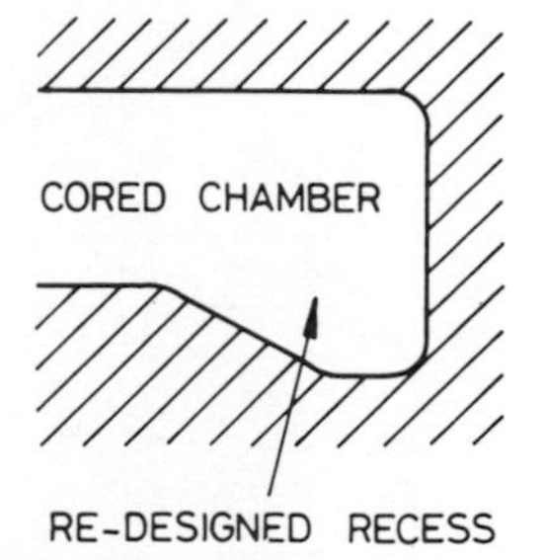

Fig. 4.42 Design to simplify sand-removal

(ii) The design of cored sections. It has already been stated that a hole or holes should be included so that the core sand can be removed, even if the cored hole must be plugged at a later stage. The internal shape of the casting must be such that the core sand does not become packed in recesses from which it would be difficult or impossible to remove it. Fig. 4.41 illustrates part of a cored section that would present difficulty when fettling, and fig. 4.42 illustrates a more acceptable design.

4.2 Design of die-castings

When a casting is produced by die-casting it is introduced into a metal mould (or *die*). In the pressure die-casting process (also known simply as *die-casting*) a machine is used to force the metal into the die, but in the gravity die-casting process (also known as *permanent mould casting*) it is poured into the die. In this respect the latter resembles sand-casting. The basic design requirements for die-casting are the same as those for sand-casting but there are special requirements associated with the use of metallic moulds instead of a sand mould. The comments in this section refer mainly to the design of pressure die-castings; the design of gravity die-castings presents fewer problems and is discussed at the end of this section.

The problem areas associated with pressure die-casting can be identified as:

(a) the removal of the finished casting from the die;

(b) the behaviour of the metal during injection, solidification, and cooling in the solid state;

(c) the manufacture and the working life of the die;

(d) the fettling of the casting.

Unlike those associated with sand-casting, these problem areas are not considered in operation sequence because the design 'priority order' does not follow the operation sequence. In addition to solving the basic problems the design must be such that full advantage is taken of the ability of die-casting to produce thin walls, complicated shapes and intricate detail.

(a) Design to enable the finished casting to be removed from the die

Die-casting presents problems that are different from those associated with sand-casting. The main problem is the restriction of shape caused by the use of a metal mould that cannot be broken to extract the finished casting. Except in certain specialised plant, the die is in two parts; if the casting is basin-shaped, the part of the die set that produces the inside is often termed the core. Holes whose axes lie in the direction of die-opening movement can be produced by cores that project from the dies; but holes in other directions would require retractable cores to allow the casting to be removed. Hollow, box-like shapes cannot be produced because they would require the use of sand cores, which would be broken and washed away by the pressure of the molten metal during its injection.

If necessary the joint line between the two members can be cranked, or even curved, enabling an awkward casting to be removed from the die or a special shape to be produced, but it must be emphasised that a die with a plane joint face is easier and cheaper to produce, and gives greater accuracy.

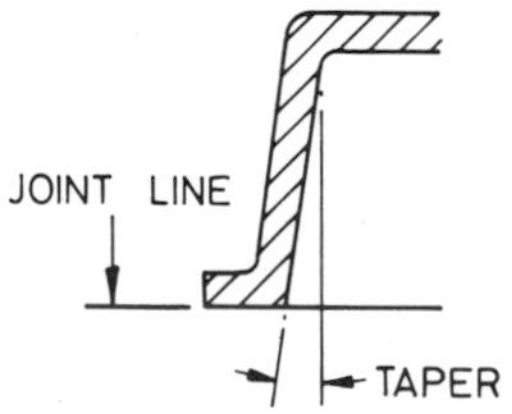

JOINT LINE
TAPER

Fig. 4.43 Location of taper

(i) Tapers. Surfaces that would lie exactly in the direction in which the dies move to release the casting should be inclined at an angle known as the *taper* (similar to the draw angle in sand-casting) so that the casting can be rapidly removed from the die without scuffing (see fig. 4.43).

(ii) Re-entrant shapes. Castings should be designed without re-entrant shapes. For example, the casting shown in fig. 4.44 has an internal flange so that the casting has a clean external appearance. But this could only be produced by using a collapsable core. The version shown in fig. 4.45 has a less clean external appearance but it can be easily produced.

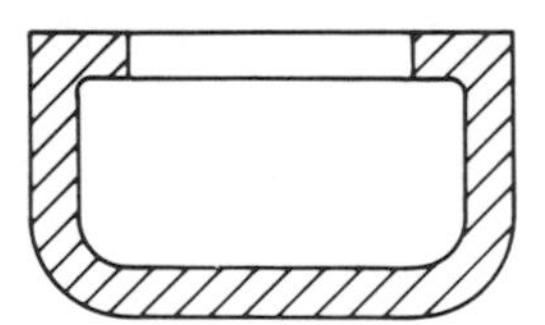

Fig. 4.44 Design that requires a collapsable core

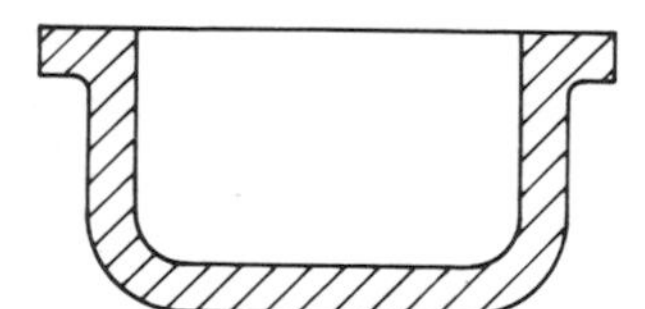

Fig. 4.45 Design that does not require a collapsable core

Careful design can often eliminate re-entrant shapes. For example, the shape shown in fig. 4.46 includes a re-entrant feature (the flange on the right); the problem is eliminated by re-designing the internal shape and having an inclined joint face (one that is not at a right angle to the direction of core withdrawal) as illustrated by fig. 4.47. Fig. 4.48 shows another

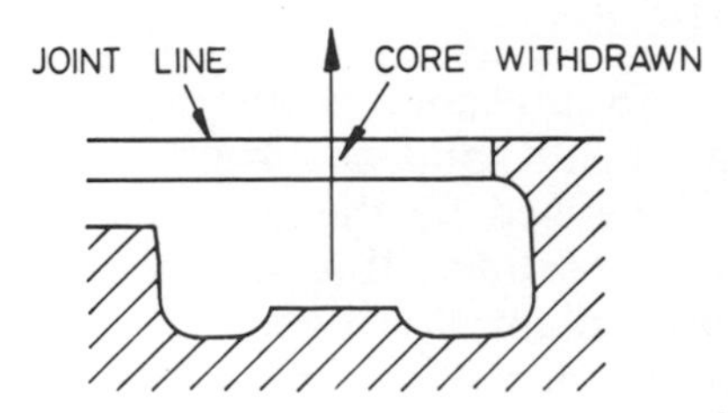

Fig. 4.46 Design with a re-entrant feature

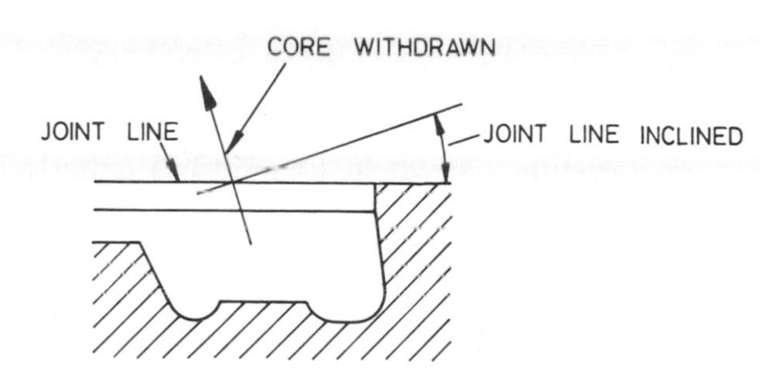

Fig. 4.47 Improved design that eliminates re-entrant feature

example of re-design to eliminate re-entrant features. Version A has two re-entrant features (the two ribs at 45° to the joint line and part of the profile). These are modified in version B to eliminate the re-entrant effect.

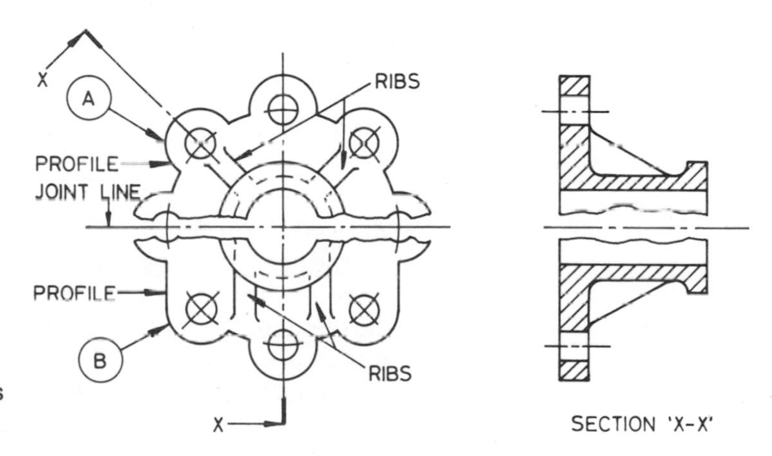

Fig. 4.48 Re-entrant features and their elimination

(iii) Bosses and re-entrant effects. Fig. 4.49 shows a box-like casting with bosses that are placed on the inside, to produce a clean external shape, but which constitute re-entrant features demanding a collapsable core. By placing the bosses on the outside of the casting, as shown in fig. 4.50, there will be no core problems but the external shape will be less clean. The version

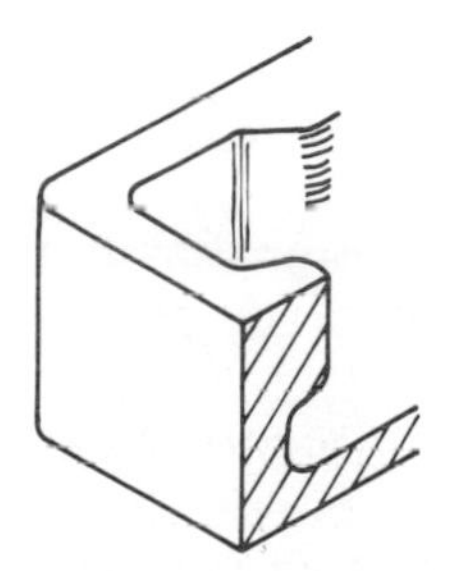

Fig. 4.49 Internal bosses

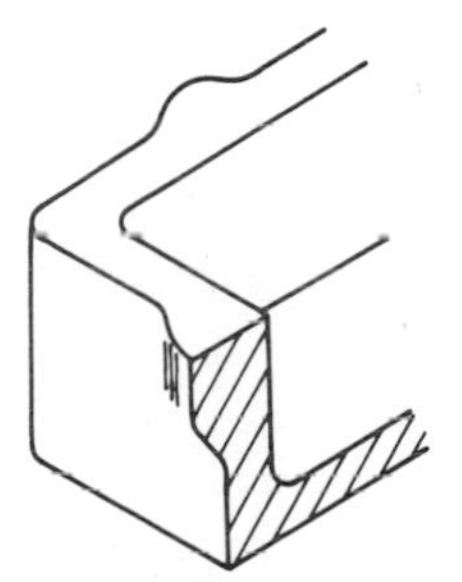

Fig. 4.50 External bosses

shown in fig. 4.51 gives a clean external shape, combined with an internal shape with no re-entrant problems, but the local heavy section at A will cause a poor structure to be produced. The version shown in fig. 4.52 is a fairly common compromise in which the external shape is reasonably clean and the re-entrant feature is eliminated.

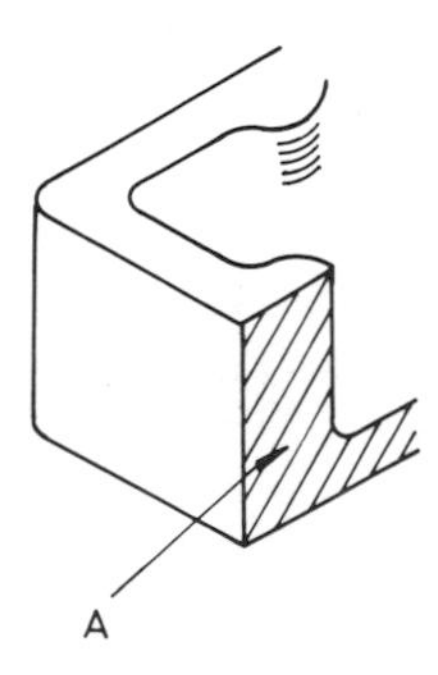

Fig. 4.51 Design that causes a heavy-section effect

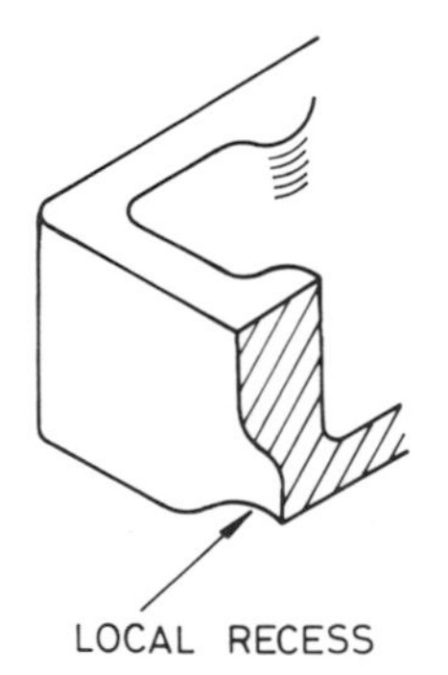

Fig. 4.52 Acceptable compromise

(iv) Cores related to joint line. When a casting with deep recesses is produced by dies with long projections that form the cores, the design should be such that the long cores only project from one die member, so that the casting can be carefully removed from the intricate member when the plain one is removed. Fig. 4.53 shows a casting that requires long cores that project from both die members and fig. 4.54 shows an improved version.

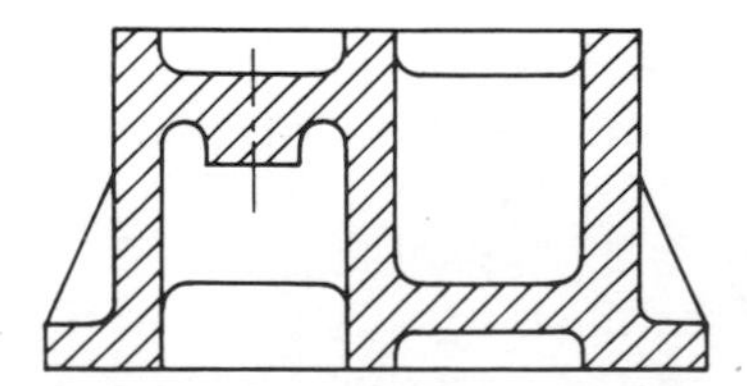

Fig. 4.53 Design with awkward recesses

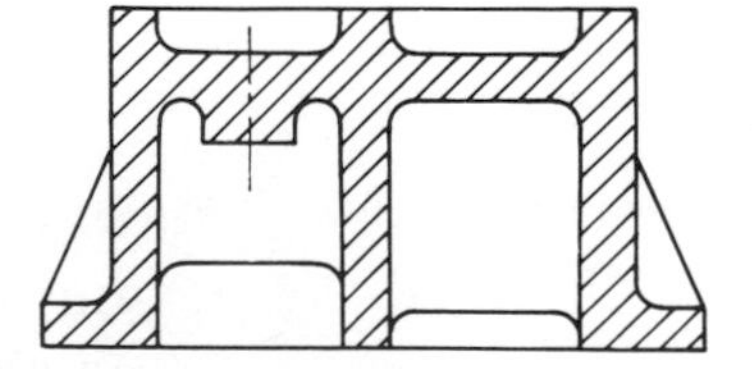

Fig. 4.54 Improved design

(v) Design to avoid retracting cores. Cored features that lie at an angle to the direction of die movement and of casting ejection require retractable cores that increase the casting time and tend to cause casting inaccuracy as a result of the wear caused by their movement.

Very often, by slight modification, an acceptable version of the desired shape can be produced without the use of retracting cores, as illustrated by the following example. The basic requirement is a box-like casting with side 'windows' (see the part section, fig. 4.55, showing two of these windows) that require retracting cores. Fig. 4.56 illustrates one variation of the basic design that can be produced by localised projections from the die.

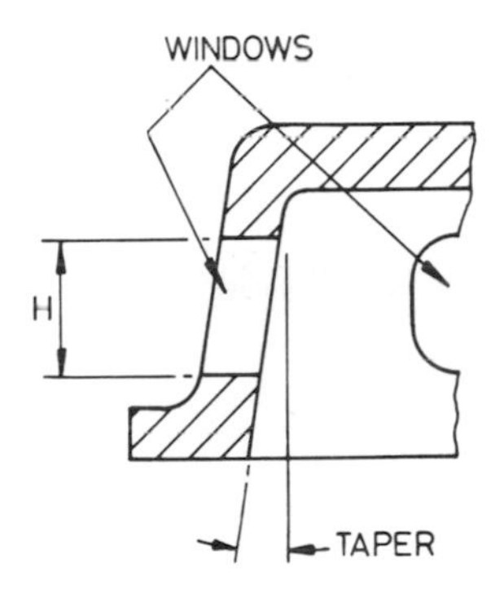

Fig. 4.55 Design that requires retractable cores

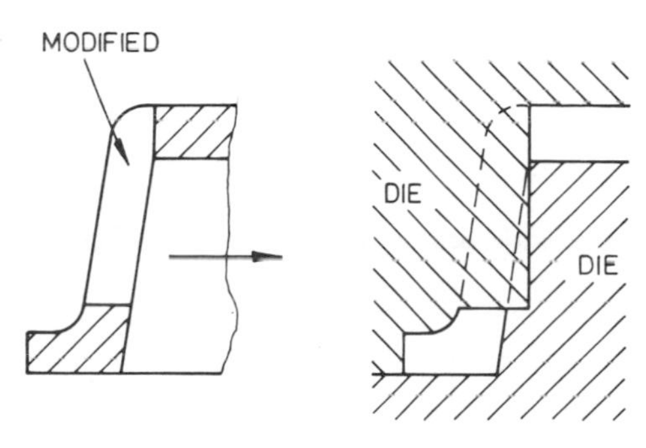

Fig. 4.56 Variation to eliminate retractable cores

Fig. 4.57 shows another variation in which, by combining an increase in the taper angle with an increase in the window height H, gives a shape that is nearer to the basic requirement and can be produced without retracting cores, using the die set shown in fig. 4.58. When retracting cores are unavoidable the casting design must be such that they do not become trapped within the casting.

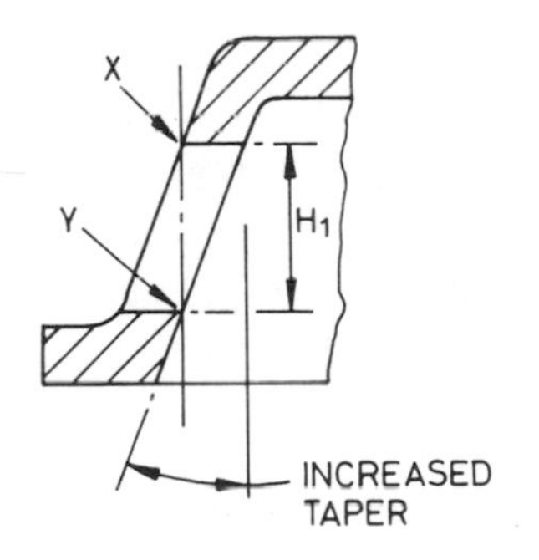

Fig. 4.57 Modification of initial design

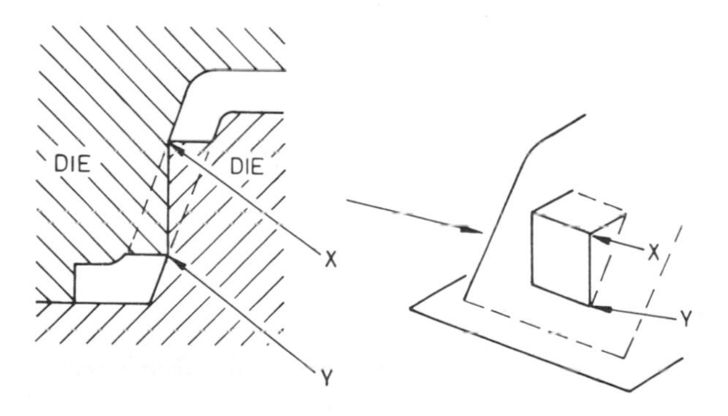

Fig. 4.58 Die-set to produce the modified design

Fig. 4.59 shows an elbow piece that could be produced by a sand core (sand cores can be used when gravity die-casting) but could not be produced by the metal cores that are necessary when pressure die-casting because of the shape at A. Fig. 4.60 shows how the inside shape can be modified to enable the cores to be withdrawn from the finished casting.

(vi) Core support. When a long slender core is used to produce holes (as shown in fig. 4.61) it should, if possible, be supported by the mating die so that it does not deflect during the injection of

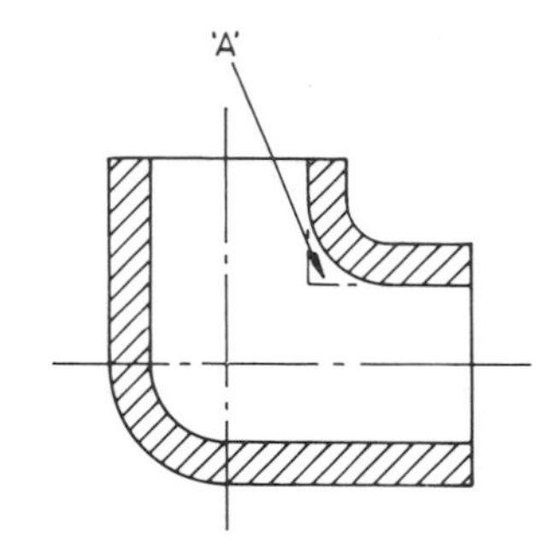

Fig. 4.59 Incorrectly-designed elbow piece

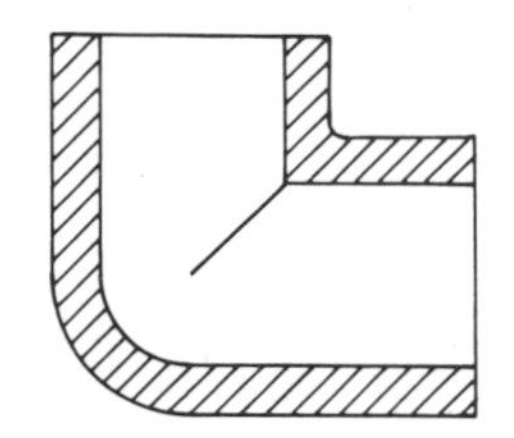

Fig. 4.60 Correctly-designed elbow piece

the metal, causing inaccuracy or making it difficult to remove the finished casting. This can only be done by making the hole a 'through hole' as shown in fig. 4.62.

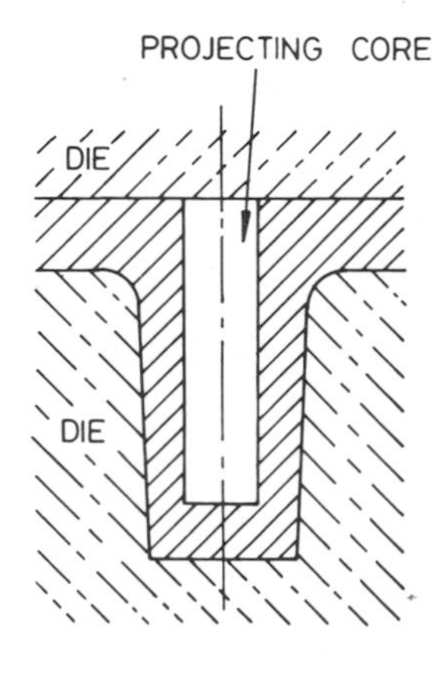

Fig. 4.61 Design that requires an unsupported core

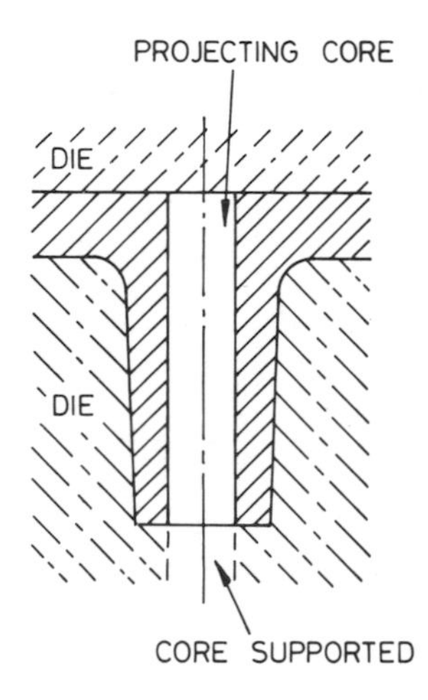

Fig. 4.62 Design that allows core to be supported

(b) Design to control the behaviour of the metal during the die-casting process

The effect of heavy sections, variation of casting thickness, hot spots etc., which were discussed with reference to sand-casting (section 4.1(b)), must be considered when die-castings are designed. The ability of the die-casting process to produce more complicated shapes often tempts the designer to produce designs that are of poor structural quality, as illustrated in fig. 4.63 in which the region at A is much thinner than the rest of the casting and becomes brittle as a result of increased local hardness following the rapid cooling. A slight modification, producing the version shown in fig. 4.64, will cause the metal to cool less rapidly and so prevent it from becoming brittle.

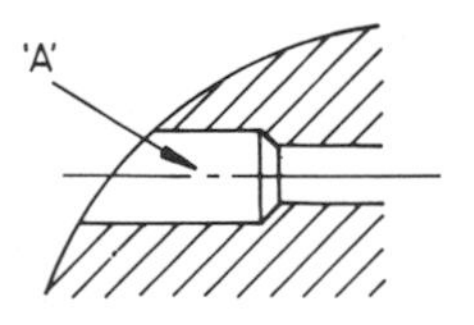

Fig. 4.63 Design that causes local brittleness

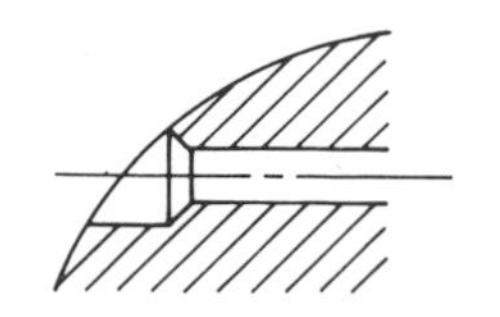

Fig. 4.64 Improved design

The ability of the process to produce thin castings may result in the design of castings that distort after solidification has taken place. This can be overcome by careful ribbing, as shown in fig. 4.65. The ribbing can often be made a decorative as well as functional feature of the casting.

(c) Design to simplify the manufacture of the die and to extend its life

The manufacture of the die (die-sinking) is an expensive operation. The casting design must be such that the die can be

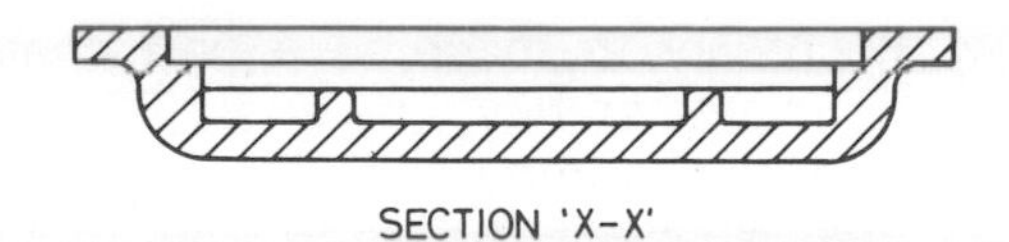

X X

Fig. 4.65 Ribs to prevent distortion

produced as accurately as is required at minimum cost. Also the design should not include features such as knife edges which wear rapidly.

(i) Corner radii. Small external corner radii on the casting are produced by the corner radii in the die cavity which, when conventional die-sinking methods are used, are formed by the die-sinking cutter. Radii such as R_1 and R_2 in fig. 4.66 should be large enough to be formed by a cutter of an acceptable diameter. Care must be taken to avoid having to change the cutter during the machining of a recess in the die because it is almost impossible to resume the machining without producing a step. Radii such as R_1 (produced in the die by the cylindrical 'face' of the cutter) should therefore be of the same size, as should radii such as R_2 (produced in the die by the end of the cutter).

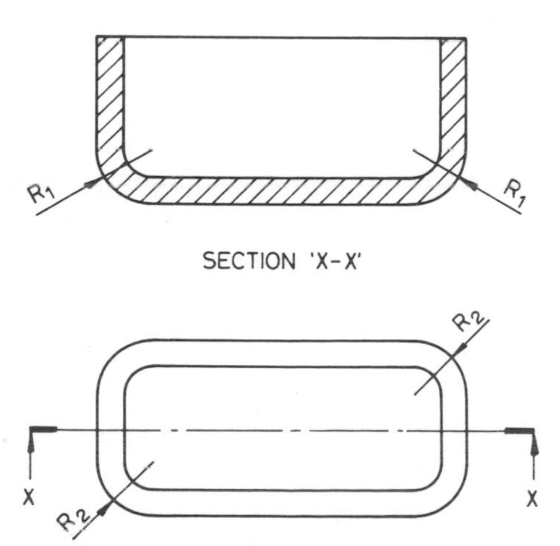

Fig. 4.66 Corner radii related to die-sinking

(ii) Casting shape to suit the die-sinking operation. When designing a die-casting the designer must think in terms of the die-sinking operation rather than the operations that would be

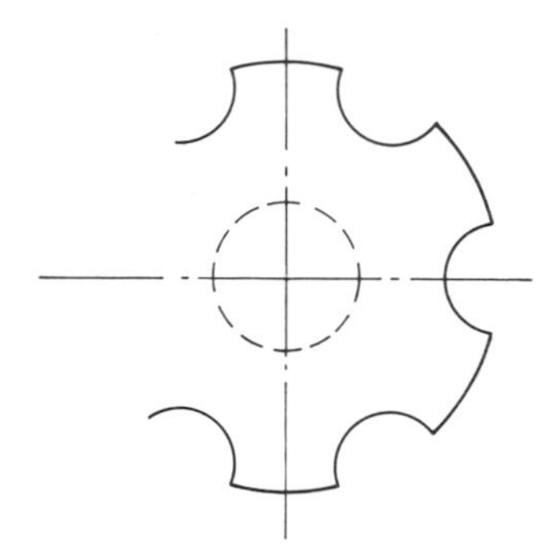

Fig. 4.67 Finger-grip design to suit machining

necessary to machine a part shaped like the casting. This is illustrated by figs 4.67 and 4.68 which show two designs for the finger grips in a knob. The grips shown in fig. 4.67 would be of ideal shape if the knob was to be machined from bar stock, because it could be turned to the outside diameter and then a milling cutter could be fed radially inwards to produce each of the recesses that form the finger grips. When these grips are produced by casting the die will be as shown in fig. 4.69; but the recesses in the die cannot be produced by feeding the die-sinking cutter radially outwards because of their shape. The grips shown in fig. 4.68 are of a shape that would be unsuitable if the knob was to be machined from bar stock; but when the grips are produced by casting the die will be as shown in fig. 4.70 and the recesses in the die can be produced by feeding the die-sinking cutter radially outwards.

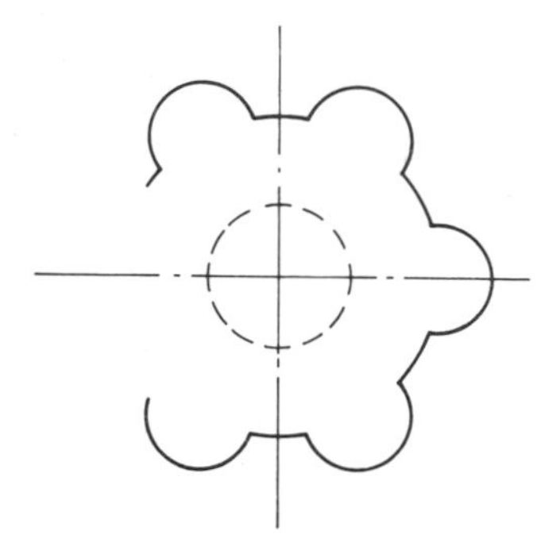

Fig. 4.68 Finger-grip design to suit die-casting

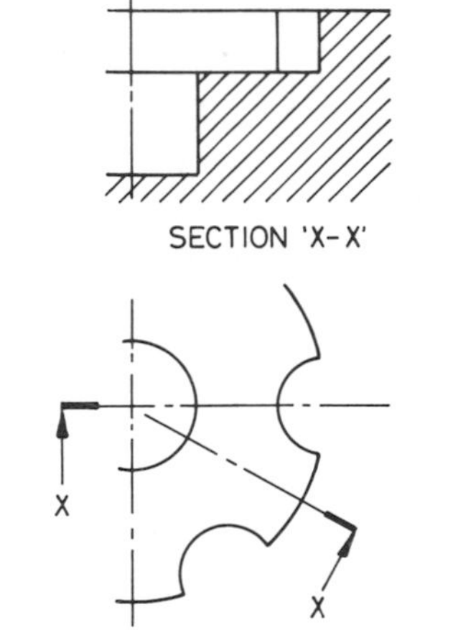

Fig. 4.69 Die recesses to suit 4.67

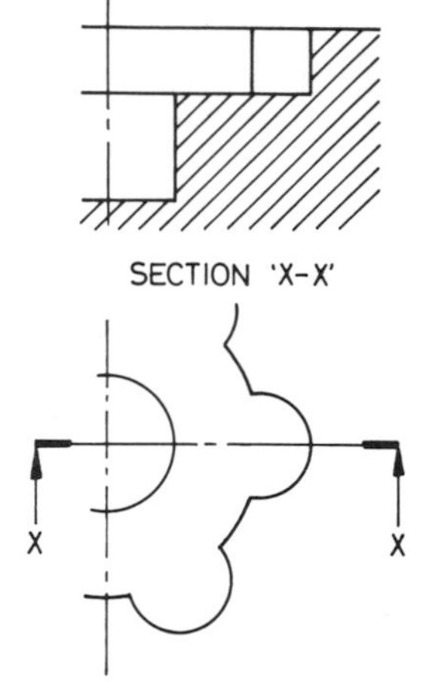

Fig. 4.70 Die recesses to suit 4.68

(iii) Casting shape to minimise die wear. The casting should not include knife edges and similar features that, in addition to making it difficult to machine the die, will promote die wear and so reduce its working life. Fig. 4.71 illustrates two forms of fluted surface. Version A requires a die with knife edges between the

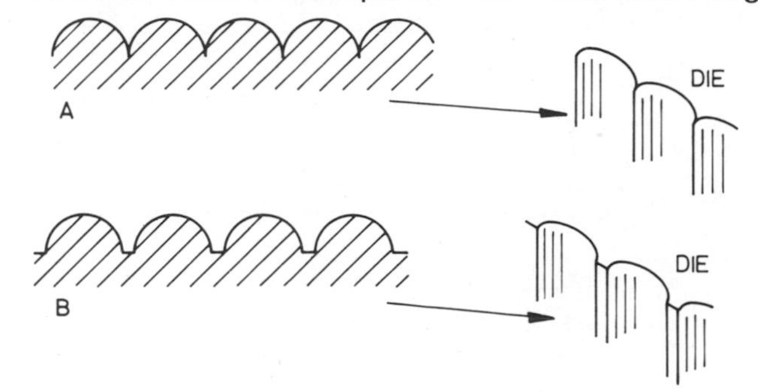

Fig. 4.71 Alternative forms of flutes

recesses; it is not easy to machine the recesses to produce a crisp shape and the knife edges will soon wear, producing a poor casting. The die to produce version B will not have knife edges and will therefore be easier to produce and will have a longer working life.

(d) Design to assist fettling

The designer must ensure that the ability of the die-casting process to produce an intricate casting does not lead to an awkward fettling operation. For example, in fig. 4.72, profile A is awkward to fettle, but by simplifying it to obtain profile B the casting becomes less awkward to fettle and therefore less expensive. Intricate detail may become blurred during fettling unless care is taken at the design stage to separate the areas to be fettled from those with intricate detail.

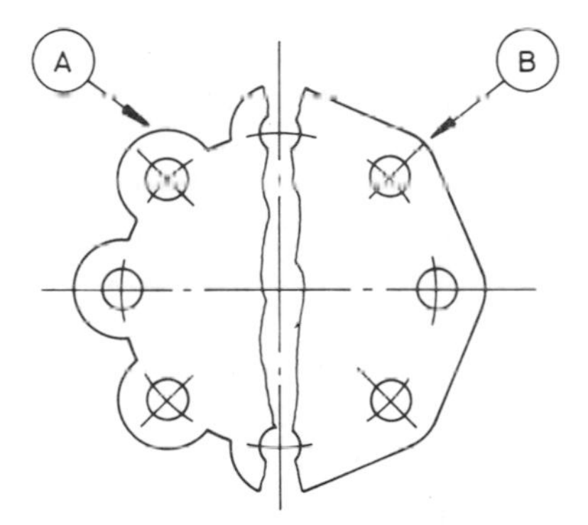

Fig. 4.72 Profiles related to fettling

(e) The design of inserts

The incorporation of inserts, such as studs and bushes, to produce local wear resistance or strength is a feature of die-casting. To be effective these inserts must be stronger, or more suitable for their purpose than the casting metal, and be made of a material with a higher melting point than that of the casting. Their shape must be such that they are gripped by the casting material when it is solid, so that they are not pulled out during assembly and cannot move when in service. Fig. 4.73 shows four ways in which a stud can be retained against the action of the nut and spanner at the assembly stage, when the stud tends to be rotated and pulled out. Fig. 4.73(a) shows a simple stud with a coarse diamond knurl that retains it; the shoulder of the cheese head in Fig. 4.73(b) prevents the stud from being pulled out and the screwdriver slot prevents it from being rotated. The 'mal-formed' head (like a wing nut) in fig. 4.73(c) effectively anchors in the stud, but a forming operation is necessary to produce it. The slotted stud shown in fig. 4.73(d) is commonly used; the two slots preventing both rotation and axial movement.

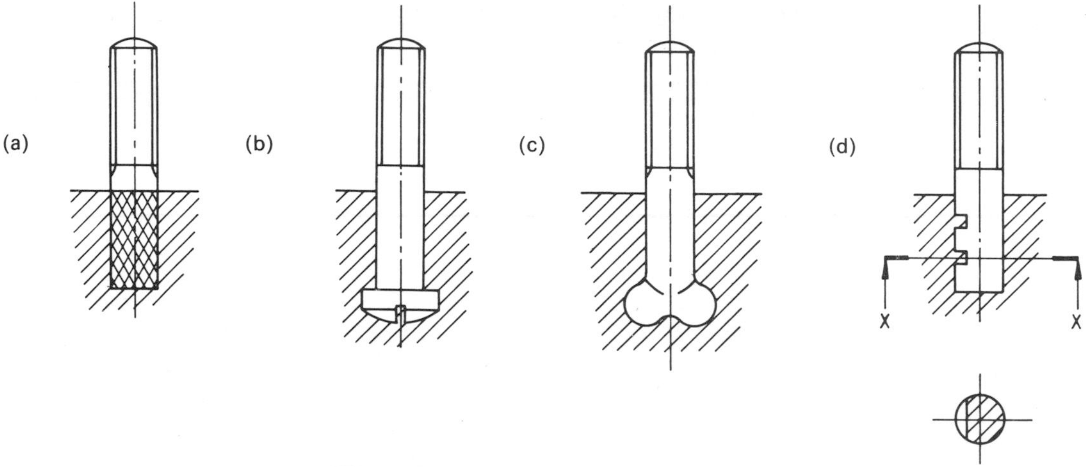

Fig. 4.73 Designs for stud-retention

SECTION THROUGH STUD AT 'X–X'

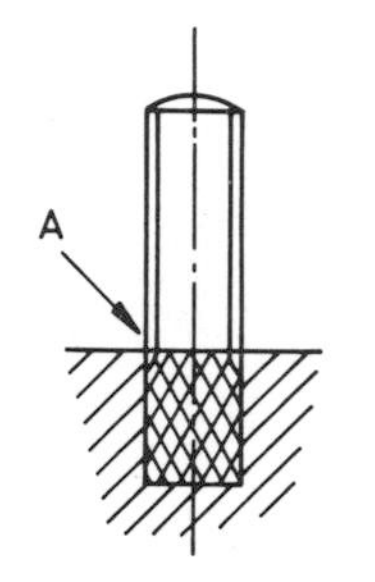

Fig. 4.74(a) Incorrectly-designed stud

It will be seen that in all these examples the threaded portion does not extend down to the surface of the casting. This is because if the threads extended to the surface the casting metal would run up the threads at A (see fig. 4.74(a)) and it would be almost impossible to remove it, even by running a die down the threads. Fig. 4.74(b) shows the improved design.

When a bearing is introduced into a through hole, its design must be such that rotation and axial movement in both directions is prevented. Fig. 4.75 shows how this is achieved using a coarse diamond knurl, and fig. 4.76 shows how this is achieved with two flats. (The latter is similar to the system used for studs, as shown in fig. 4.73(d).) When a bearing or similar insert is introduced into a through hole, its axial position should not be allowed to change during casting. The design shown in fig. 4.77 is not good because although the insert is located on the projecting core it is only constrained axially in one direction during casting. If the through hole is stepped, as shown in fig. 4.78, the core can be split as shown so that a core projects from each of the die members. The insert can now be located over one of the projecting cores and constrained axially by the combined effect of the two parts of the die.

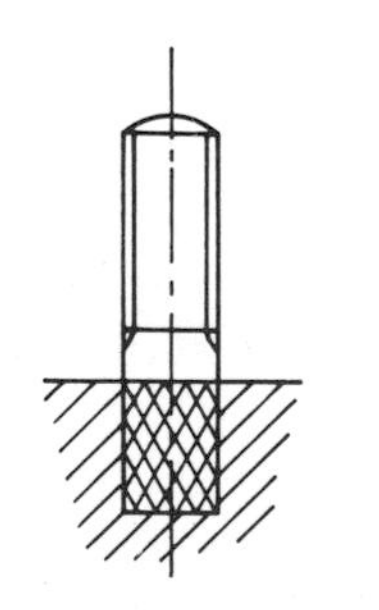

Fig. 4.74(b) Correctly-designed stud

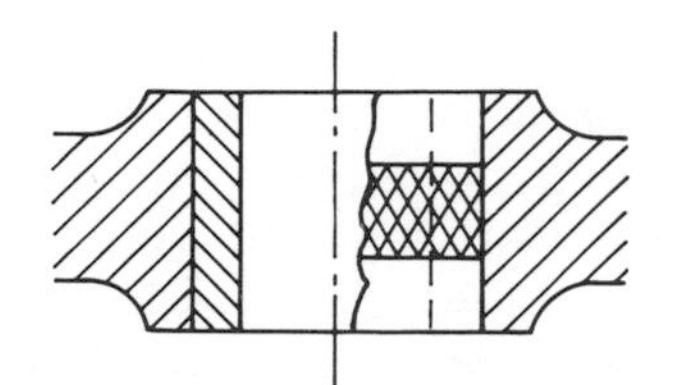

Fig. 4.75 Design for bearing retention

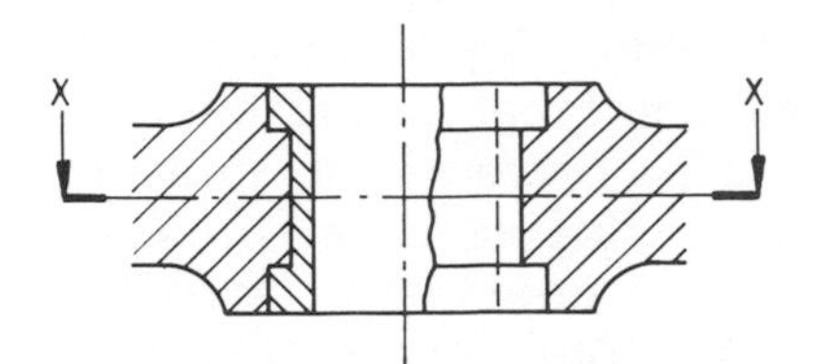

Fig. 4.76 Design for bearing retention

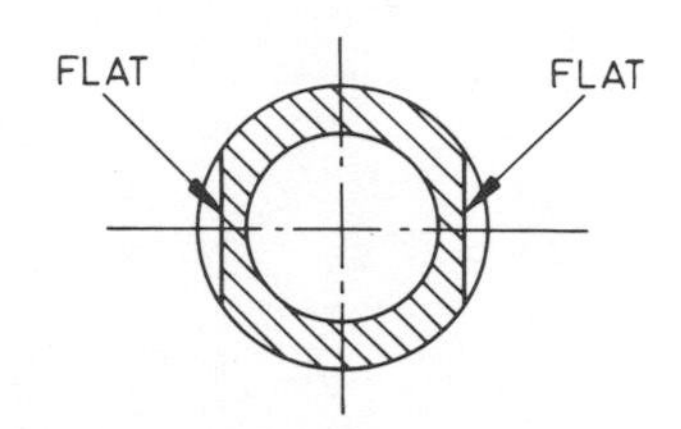

SECTION THROUGH INSERT AT 'X–X'

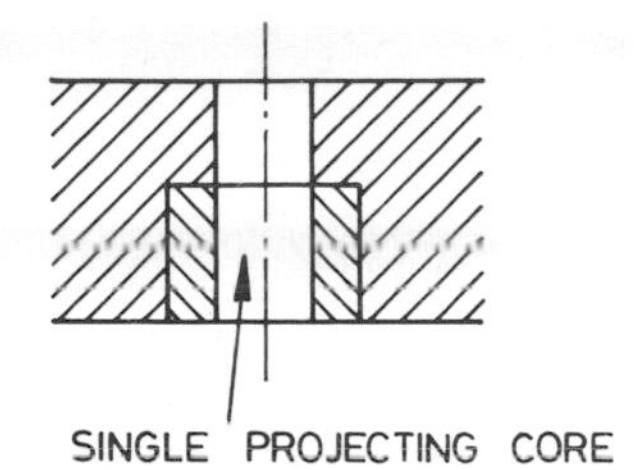

Fig. 4.77 Insert that is not fully constrained during casting

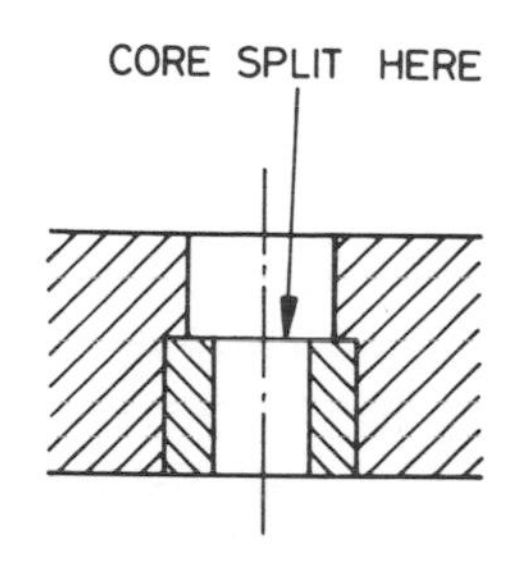

Fig. 4.78 Insert that is fully constrained during casting

(f) Lettering

Features such as lettering and decoration can be incorporated without difficulty provided that care is taken regarding shape and position. The features should be raised and not incised because to produce raised features it is only necessary to carry out the simple operation of incising the die. To produce features that are incised in the casting, a lengthy operation must be carried out to remove metal from the face of the die, leaving locally raised features.

Lettering and decoration must be positioned where they do not produce re-entrant problems. Ideally they should be positioned on the joint face, or on one that is parallel to it. When it is necessary to position these features on faces that are inclined, as shown in fig. 4.79, they must be shaped so that their sides are inclined to enable the casting to be rapidly removed. Lettering and similar features should be positioned so that they do not become blurred when the casting is fettled or polished.

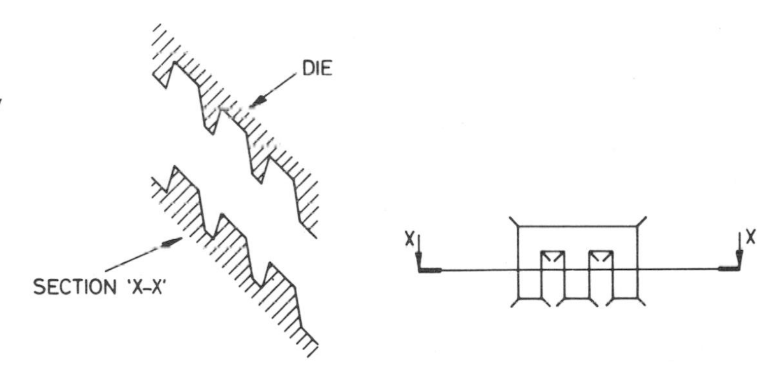

Fig. 4.79 Lettering on an inclined surface

(g) Accuracy

The die-casting process produces a high degree of dimensional and surface accuracy. Table 4.1 shows typical tolerances for critical dimensions and table 4.2 shows typical tolerances for non-critical dimensions.

When the joint line affects a dimension, as shown in fig. 4.80, the tolerance for it must be larger than that shown in tables 4.1 and 4.2. This additional *joint line tolerance* depends upon the

Table 4.1 Tolerances for critical dimensions

Size	Die-casting alloy	
	Zinc alloys	Aluminium alloys and magnesium alloys
Basic tolerance up to 25 mm	± 0.05 mm	± 0.07 mm
Additional tolerance for each 25 mm of dimension	± 0.02 mm	± 0.04 mm

Table 4.2 Tolerances for non-critical dimensions

Size	Die-casting alloy	
	Zinc alloys	Aluminium alloys
Basic tolerance up to 25 mm	±0.20 mm	±0.20 mm
Additional tolerance for each 25 mm of dimension	±0.02 mm	±0.04 mm

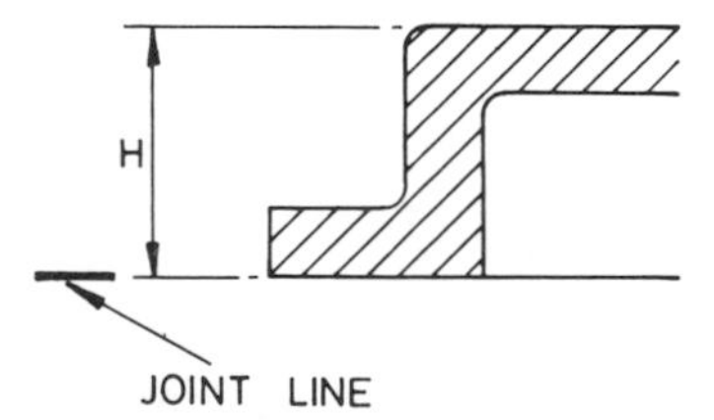

Fig. 4.80 Dimension affected by joint line

material being cast and the projected area of the die-casting at the die-parting plane. When the projected area is up to 15 000 mm^2, the additional tolerance is ±0.07 mm for zinc alloys and ±0.10 mm for aluminium alloys and magnesium alloys; it must be increased with increase in projected area.

Table 4.3 shows typical flatness tolerances, the dimension being the maximum dimension (diameter of a circular surface, or diagonal of a rectangular surface).

Table 4.3 Flatness tolerance (as cast)

Size	All alloys
Basic tolerance up to 70 mm	±0.20 mm
Additional tolerance for each additional 25 mm	±0.07 mm

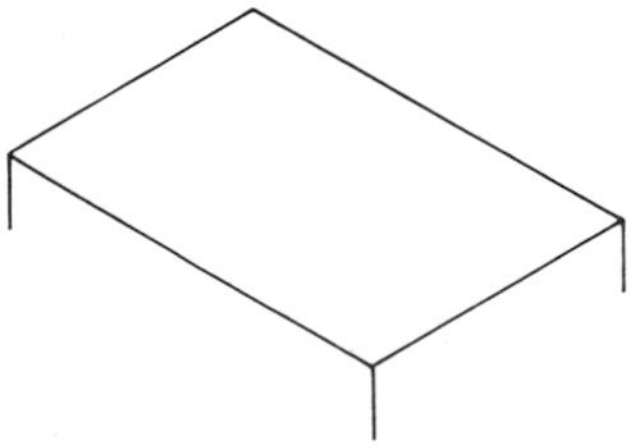

Fig. 4.81 Large plain surface

Large plain surfaces should be avoided because flatness errors will be obvious. Where possible they should be broken up with stippling or other decoration as illustrated by figs 4.81 and 4.82.

A beading at the joint line, as shown in fig. 4.83, should be avoided because a step would be produced if the dies did not match up perfectly and any attempt to improve the appearance by fettling would be time consuming and probably unsuccessful. The form shown in fig. 4.84 would not be affected by die mismatch.

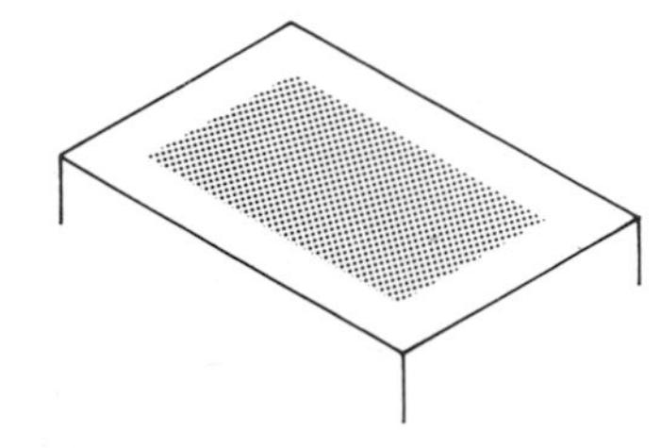

Fig. 4.82 Broken-up surface

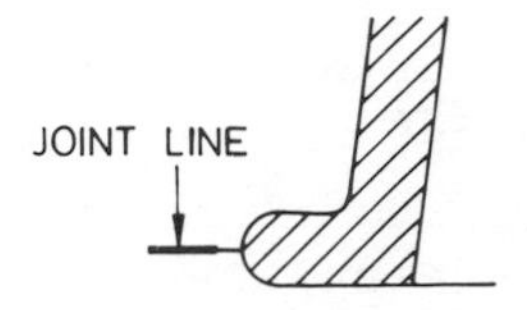

Fig. 4.83 Beading at joint line

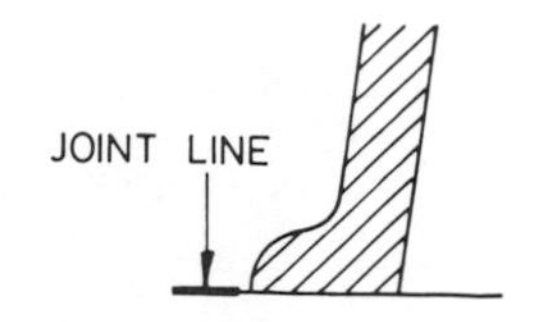

Fig. 4.84 Improved beading design

(h) Typical die-casting

Fig. 4.85 illustrates some of the points made in the preceeding sections. The lettering is on the surface that is parallel to the joint line and knurling is also cast on that surface. The cylindrical surface has an edge-milled surface that does not produce a re-entrant problem and a fettling beading is incorporated so that the edge milling does not become blurred by fettling. In this example the internal threads are machined; an alternative to machining would be to incorporate a threaded insert.

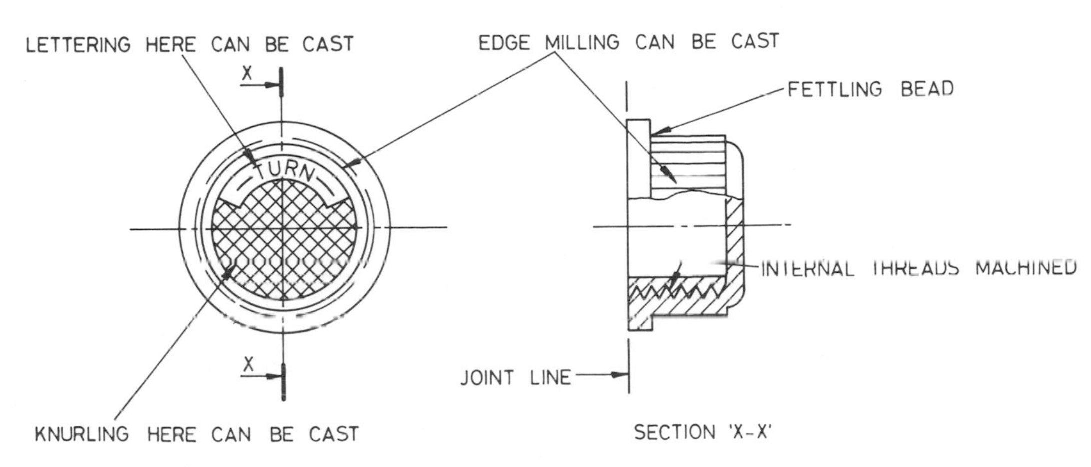

Fig. 4.85 Typical die-casting

(i) Gravity die-casting

The problems associated with pressure die-casting are also present with gravity die-casting, because a metal mould (die) is used, but they are somewhat reduced because multi-sectioned dies can be used to overcome re-entrant problems. The feeding of the molten metal under gravity alone enables separate sand cores to be used, allowing internal passage to be cast as when sand-casting. The effect of local heavy sections and hot spots, and similar problems already discussed with reference to sand-casting (section 4.1 (b)), must be given special attention, particularly since this process is so similar to sand-casting.

4.3 Design of investment-castings

Investment-casting allows greater freedom of design than do other casting processes because the pattern is removed from the mould by melting it, the mould is broken open when the metal is solid (to obtain the casting) and also because the process is capable of producing very accurate castings in a wide range of materials. This does not imply that there are no problems associated with the process, or that the material behaves in a different way during its solidification and solid state cooling than it does when other casting processes are used.

The problem areas associated with investment-casting can be identified as:

(a) the manufacture of the die that is used to produce the pattern;
(b) the manufacture of the pattern;
(c) the behaviour of the metal during pouring, solidification and cooling in the solid state;
(d) the fettling of the casting.

These problems are listed in operation sequence, but areas (a) and (b) are so closely related that they will be considered together. As in the case of die-casting, the design must be such that in addition to satisfying the basic problems, full advantage is taken of the ability of the process to produce thin sections, intricate detail and complicated shapes that may enable a single casting to replace an assembly of small components.

(a) Design to assist in the manufacture of the die and pattern

As already stated, the problems associated with the manufacture of the mould are reduced by using an expendable pattern that is melted and removed from the mould cavity. A pattern is therefore required for every casting that is produced. The pattern is produced using a die that is removed in sections from about the finished pattern; this method enables re-entrant and similar complicated shapes to be produced. The manufacture of the pattern is not, however, without its problems. A casting with local re-entrant features may require a die that consists of many small sections so that it can be removed from the pattern. It may be impossible to design a die that can be removed from all parts of the pattern. Even when it can be done, the complicated die assembly and die section removal from the pattern would make the die expensive to manufacture, cause its working life to be reduced as a result of rapid loss of accuracy caused by wear of the location pins and faces needed for the many die sections, and cause the pattern manufacturing operation to be lengthy. The designer should keep re-entrant features to a minimum and avoid undercuts at the surface of the casting, because they would be local re-entrant features and unnecessarily complicate what may otherwise be a comparatively simple die.

(b) The behaviour of the metal during the investment-casting process

The effect of heavy sections, variation of casting thickness, hot spots etc., discussed with reference to sand-casting (section

4.1(b)), must be considered when designing investment-castings. This is of particular importance because the complicated shapes that can be produced by this process often cause variation of casting structure.

(c) Design to assist in the fettling of the casting

The relationship between the location of the feeding gate and the ease of the fettling operation must be considered and, as in the case of die-casting, a fettling beading should be provided to ensure that intricate detail does not become blurred during fettling.

(d) Accuracy

Investment-casting produces a good 'as cast' accuracy in addition to its ability to allow awkward shapes to be cast. The following are typical tolerances, but in special cases one, or possibly two, of the dimensions may be produced more accurately, although this increases the tooling and operation costs.

General tolerances are usually stated as ± 0.1 mm per 25 mm but in special cases they may be reduced to ± 0.07 mm on dimensions less than 5 mm.

Axial straightness may be a problem because variation of the cooling rates within a casting causes some distortion. This can be minimised by careful casting design. Generally an axial straightness of ± 0.1 mm per 25 mm as cast is specified where straightness is a functional requirement.

Flatness is associated with distortion during cooling and a slightly dished shape can be expected over a surface. The amount of the dishing varies slightly according to the section thickness but is usually of the order of 0.2 mm.

Angles between sections may be inaccurate and the squareness of the faces within a casting may be poor, depending upon the location of the angle and the design of the casting. When the sections are well supported by ribs etc. the accuracy may be $\pm 0.5°$, but where distortion would be expected to occur a tolerance of $\pm 2°$ may need to be allowed.

5 Design for Working

The factors that influence the selection of the manufacturing method are discussed in chapter 3. In this chapter the effect of working on the material is examined and the problem areas associated with the principal working processes are identified. The detail design of wrought products is related both to the effect of working and to the need to avoid the associated problems.

Working is defined as *manipulation of metal in the solid state* after it has been cast into a suitable shape. Not all metals and alloys can be worked and alloys are usually classified either as *casting alloys* or as *working alloys* (or *wrought alloys*).

Working processes can be classified as *hot-working processes* and *cold-working processes*. Hot-working is done at a temperature that is higher than the recrystallisation temperature of the metal being worked and it does not produce residual stresses because recrystallisation causes the grains to be reformed. Cold-working is done at a temperature that is below the recrystallisation temperature of the metal being worked and is usually, but not necessarily, done at room temperature. Cold-working produces residual stresses because recrystallisation does not occur.

Most metals and alloys that can be cold-worked can also be hot-worked; but some alloys, for example α-brasses (these are brasses with less than about 37 % zinc), can be successfully cold-worked, but are unsuitable for extensive hot-working.

5.1 Hot-working

In addition to enabling the material to be deformed at a high rate, hot-working improves its strength by producing two structural changes. Firstly it develops a uniform, fine grain size because the

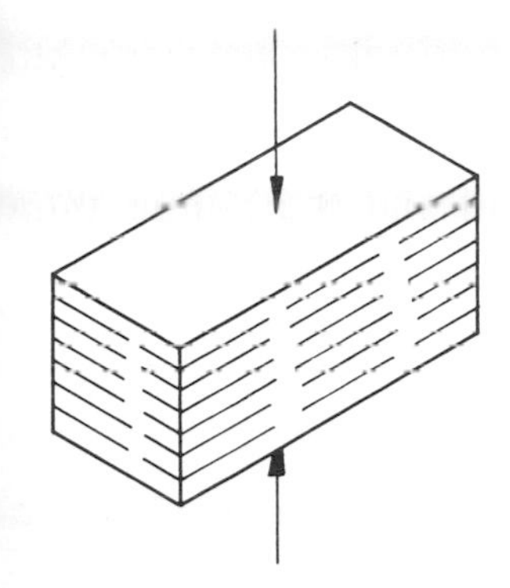

Fig. 5.1 Grain fibre developed by hot-working

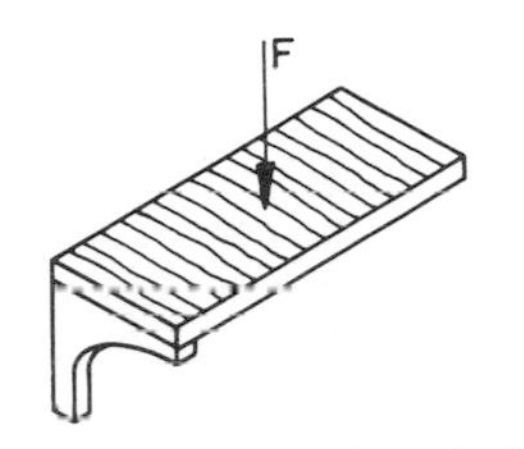

Fig. 5.2 Weak timber shelf

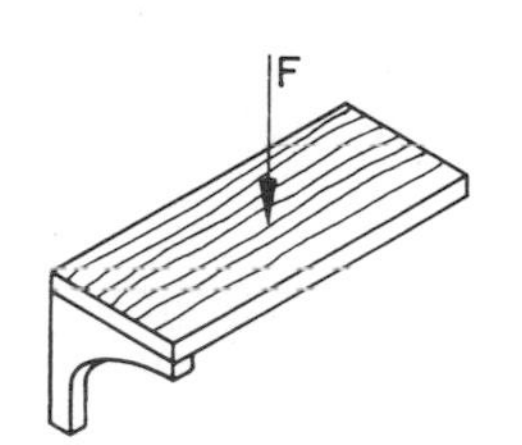

Fig. 5.3 Strong timber shelf

forces involved prevent the grains from growing after recrystallisation; with the exception of steels (whose grains can also be made fine by a heat-treatment process called *normalising*) this is the only way whereby a coarse grain structure can be made fine. Secondly it changes the shape of solid inclusions from spherical to elongated and orientates the grains to develop *grain fibre* (or *grain flow*); this gives a directional strength that can be made to coincide with the direction of the major stresses.

Fig. 5.1 illustrates the grain fibre that is developed by hot-working the material in one plane along its length. It will be seen that the grain fibre direction is at right angles to the direction of the working forces that produced it. The directional strength of a hot-worked material can be likened to that of wood. Fig. 5.2 shows a wooden shelf that is weak because the force acts along the grain, and fig. 5.3 shows a shelf that is strong because the force acts across the grain.

By working the material in the planes along its length, either in turn or simultaneously, a grain fibre as shown in fig. 5.4 and fig. 5.5 is developed. This results in an improved strength in two planes. This can be extended to develop a grain structure that

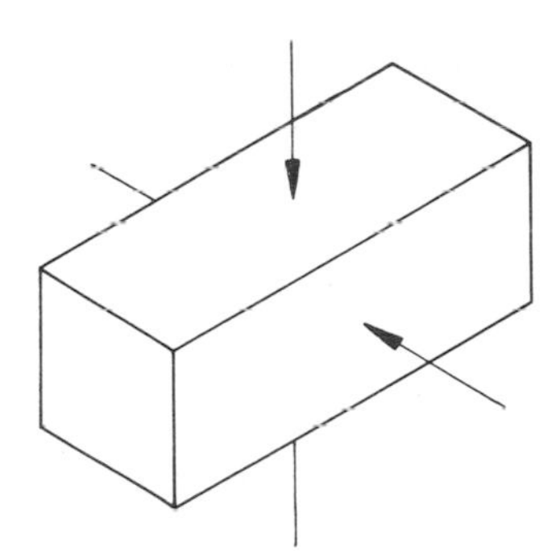

Fig. 5.4 Hot-working in two planes

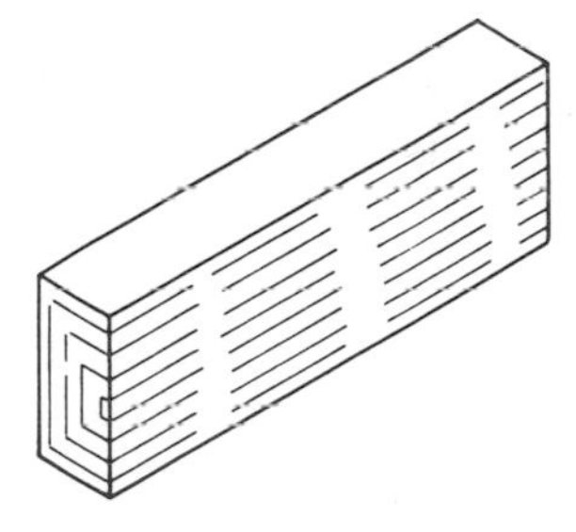

Fig. 5.5 Grain fibre developed by hot-working in two planes

resembles that of a tree trunk, as shown in fig. 5.6 and fig. 5.7, by working the material in several planes along its length. Material worked in this way has multi-directional high strength.

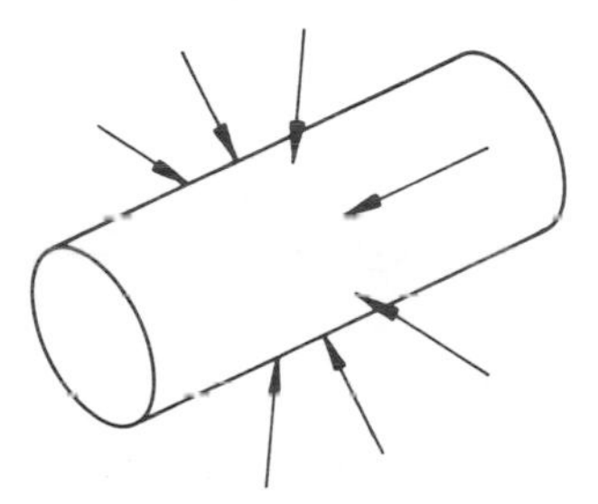

Fig. 5.6 Hot-working in several planes

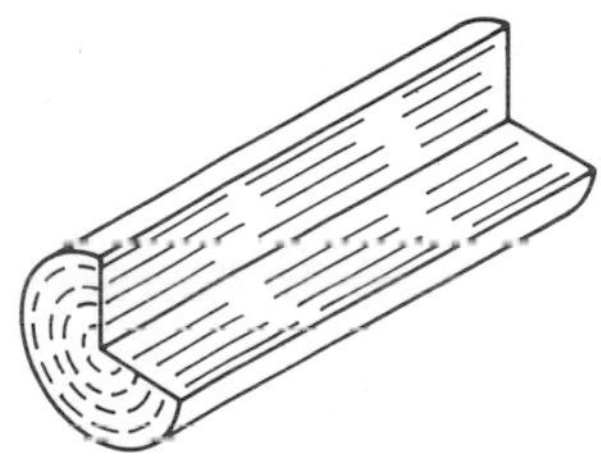

Fig. 5.7 Grain fibre developed by hot-working in several planes

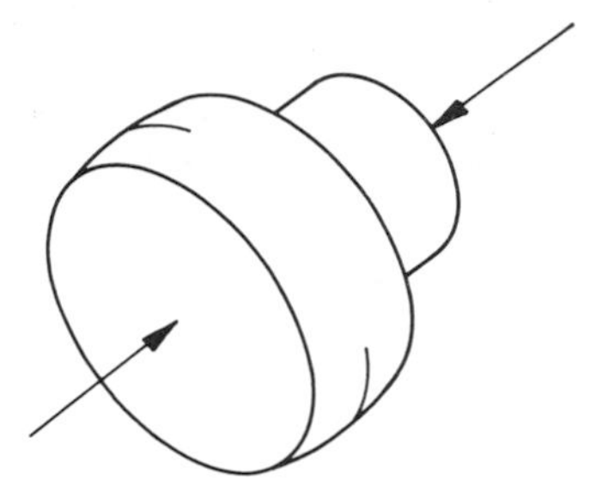

Fig. 5.8 Upsetting

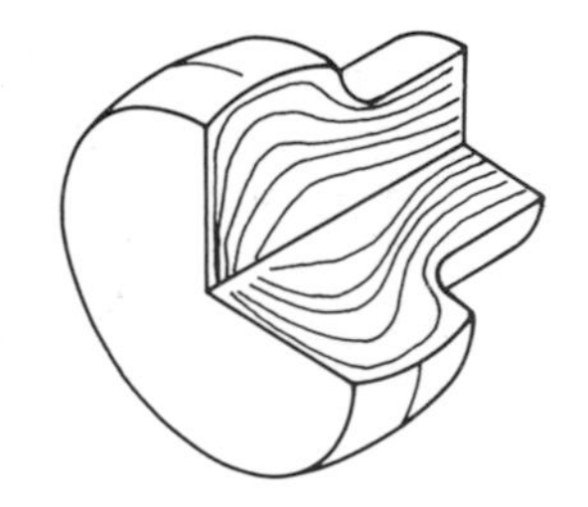

Fig. 5.9 Grain fibre developed by upsetting

When it is necessary to develop a structure with a 'best strength direction' that differs from one region of the product to another the upsetting process is used. In this process the material shown in fig. 5.6 and fig. 5.7 is axially worked and the deformation localised either by local heating or by using a die. Figs 5.8 and 5.9 illustrate how the grain fibre can be modified by this upsetting process.

The three principal hot-working processes are forging, hot-rolling and extruding. Forging produces 'single' workpieces and is therefore an alternative to casting; hot-rolling and extruding produce long lengths.

5.2 Design of forgings

Forging involves hitting the material using a hammer (manually or by machine) or squeezing it using a press. Simple tools are used when small quantities are required, and dies are used for large quantities (closed die forging, drop forging or stamping).

The problem areas associated with closed die forging can be identified as:

(a) the behaviour of the workpiece material during forging;
(b) the separation of the forging from the dies;
(c) the manufacture and life of the dies.

(a) The behaviour of the workpiece material during forging

It has already been stated that the development of the grain fibre direction so that it coincides with that of the major stress when the product is in service is an important function of hot-working. As the direction of the grain fibre is normal to that of the forces that are applied during forging it follows that the direction of the grain fibre will depend upon the shape of the die cavity. This in turn depends upon the profile of the forging. In order to obtain the required grain fibre direction the shape of the forging may differ quite considerably from that of the finished product. Forging is often an alternative to casting but the superior strength of a forged product is obtained at the expense of accuracy and complexity of shape. As stated in chapter 3, manufacture involving a forging usually involves removal of metal to obtain the required shape and accuracy. The change of grain fibre direction should not be abrupt, otherwise regions of weakness and stress concentration will be produced.

Although the separation of the forging from the dies and the manufacture and life of the dies are considered separately, these factors influence the behaviour of the material during forging because they affect the forging profile.

(b) The separation of the forging from the dies

This problem is similar to that associated with die-casting. The location of the parting line between the two sections of the die must be such that the forging will not be trapped. Also there must be no re-entrant features, and surfaces that would lie exactly in the direction of the die and forging movement must be inclined at an angle (called the *draft angle*).

The location of the parting line when forging produces more problems than when casting. This is because to accomodate the somewhat unpredictable flow of metal a greater volume of metal than is required in the finished forging is placed in the dies and the excess is allowed to flow out through the flash lane and into the gutter to form a flash that is later removed. This forms a grain fibre *outcrop* with inferior properties (see fig. 5.10). The forging design should be such that the location of the parting line ensures that the product is of a shape to suit its manufacture and has the required directional strength.

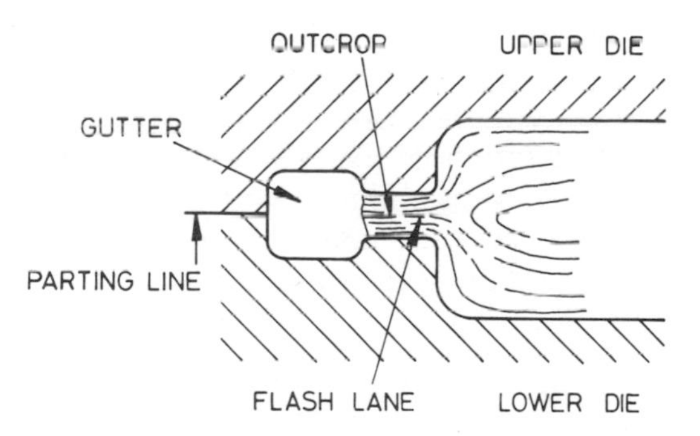

Fig. 5.10 Formation of flash

The depth of the recesses in the die into which the material must flow will affect the orientation of the forging with respect to the parting line as shown in figs 5.11 and 5.12—the aim being to minimise the depth to which the material must flow. Plane parting lines are usually employed to make die-sinking easier, but it may be necessary to locate the parting line in several planes to suit the

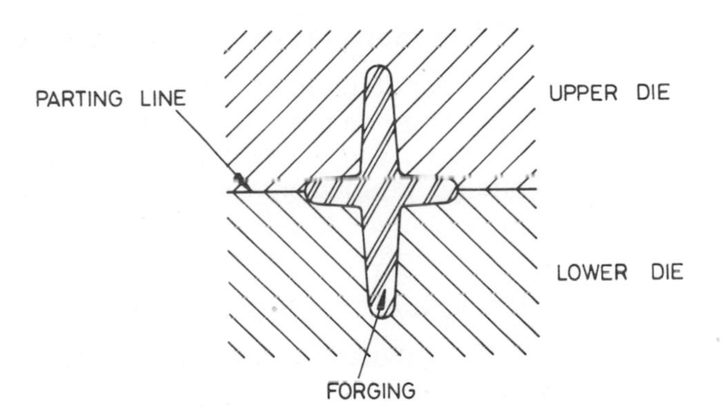

Fig. 5.11 Parting line location requiring deep die recesses

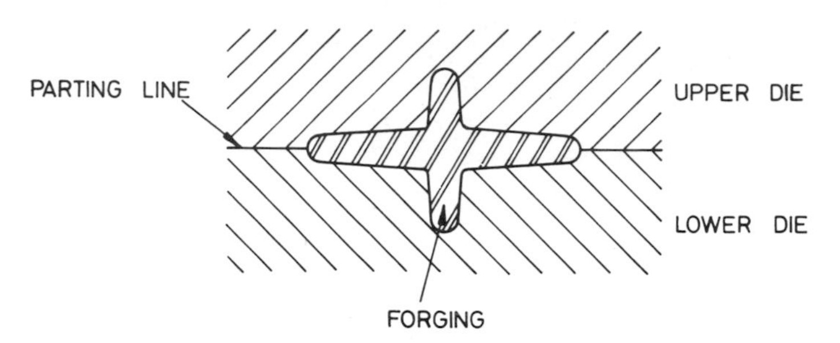

Fig. 5.12 Parting line location requiring shallow die recesses

component being produced. An alternative to employing a cranked parting line is to produce a forging with a plane parting line and to bend it afterwards.

The thickness of the flash must be large enough to ensure that the die cavities are filled before metal escapes. The flash thickness is related to the weight of the forging (it increases with forging weight) and to the thickness of webs in the forging (no webs that are thinner than the flash will be produced).

Draft angles are related to their location and to the method of die operation. Draft angles at the outside of the forging are typically 6° when a hammer is used and 4° when a press is used. Those at the inside of a forging are about 2° larger to allow for the contraction of the workpiece material (see fig. 5.13).

Generous radii should be allowed at the corners to prevent weakness (as a result of poor flow of metal during forging) and to prevent the dies from cracking; typical minimum values are shown in fig. 5.13.

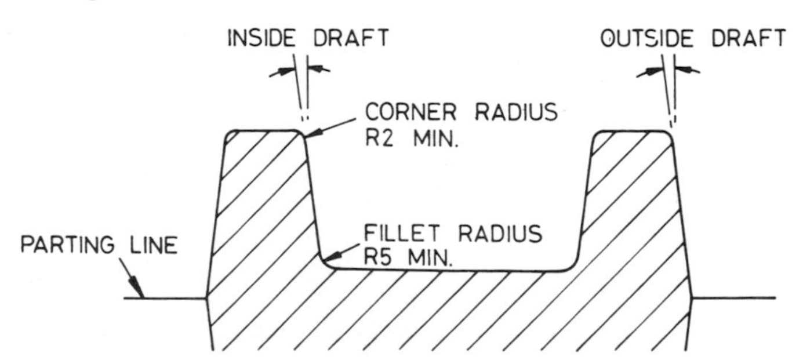

Fig. 5.13 Draft angles and radii

(c) The manufacture and life of the dies

The manufacture of the dies will, as already stated, be easier if the parting line lies in one plane and if the forging corners have large radii. Cutter changes during die-sinking, and associated blending problems, will be minimised by specifying the same corner radius and fillet radius throughout.

(d) Typical forging design

Fig. 5.14 illustrates a section through a hollow gear. Fig. 5.15 shows the ideal grain fibre—the shank should have a cylindrical

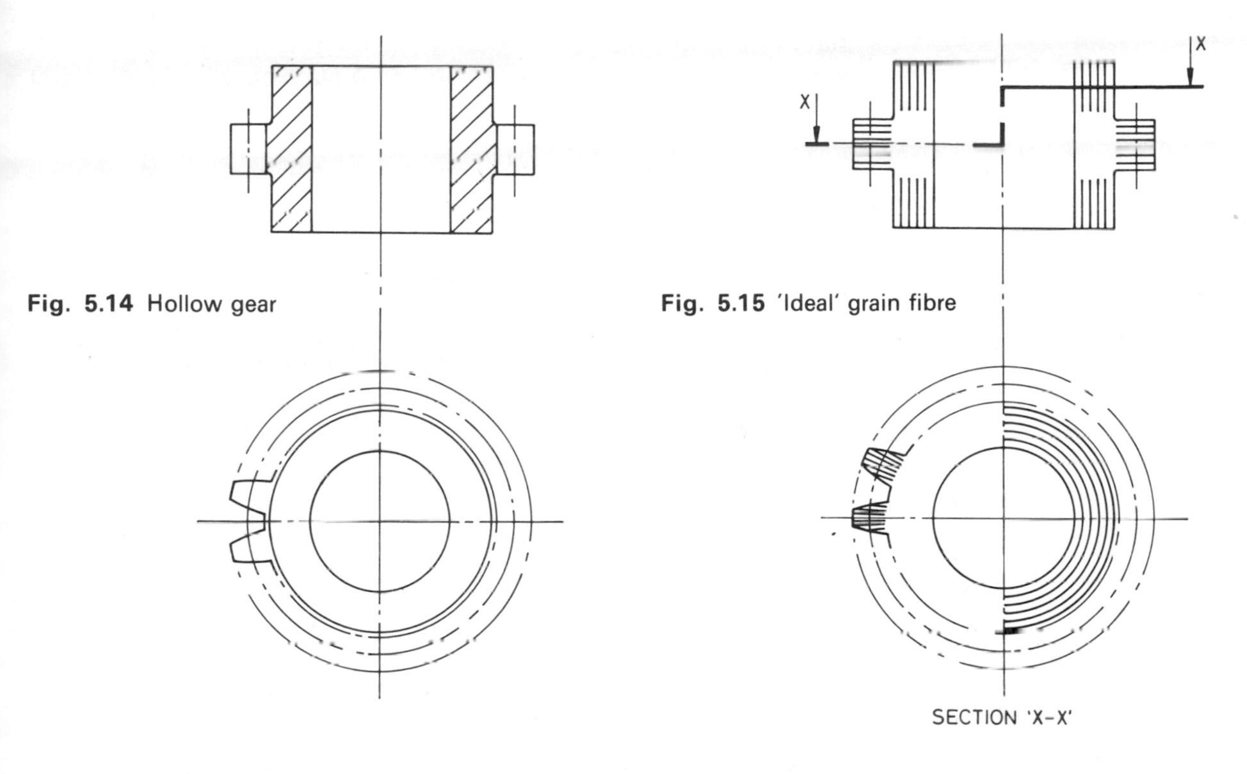

Fig. 5.14 Hollow gear

Fig. 5.15 'Ideal' grain fibre

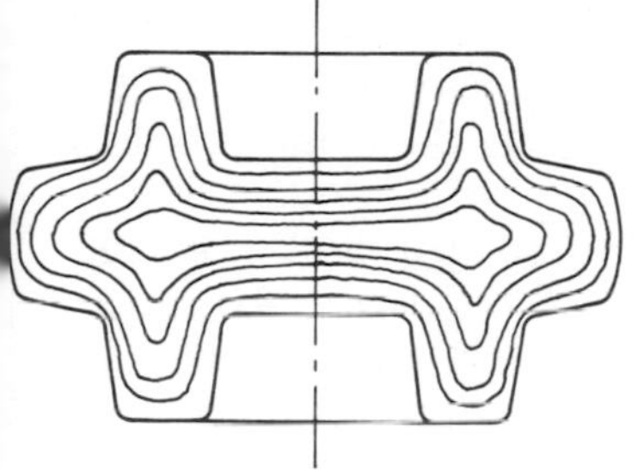

Fig. 5.16 Forging showing grain fibre

grain fibre and the teeth should have a radial grain fibre for best strenth. Fig. 5.16 shows the section through the forging for the gear, and the grain fibre obtained is indicated. By producing a forging with two recesses a grain fibre is developed that is close to the ideal one.

5.3 Design for hot-rolling

Rectangular sections are produced by passing the material through the gap between two cylindrical rollers. Complicated sections are produced using grooved rollers—the gaps formed by grooves gradually approaching the required shape and size. The problem areas associated with hot-rolling can be identified as:

(a) the behaviour of the metal during rolling;

(b) the passage of the metal through the gap or gaps between the rollers.

As with forging the products of hot-rolling are of poor dimensional accuracy and of poor surface finish, although this is not a problem when structural sections are being produced. When, as with sheet metal for a motor car body, accuracy and finish are important, the material is finished by cold-rolling.

The section shape must satisfy the assembly requirements but it is also related to the direction of the grain fibre, and therefore to

Fig. 5.17 Hot-rolled section

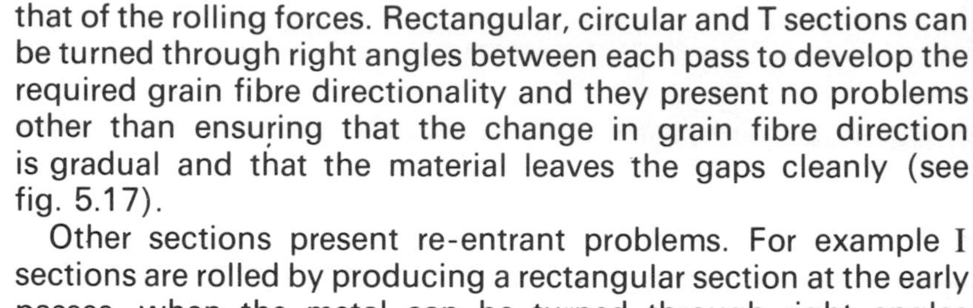

that of the rolling forces. Rectangular, circular and T sections can be turned through right angles between each pass to develop the required grain fibre directionality and they present no problems other than ensuring that the change in grain fibre direction is gradual and that the material leaves the gaps cleanly (see fig. 5.17).

Other sections present re-entrant problems. For example I sections are rolled by producing a rectangular section at the early passes, when the metal can be turned through right angles between each pass. When, at the later passes, the I shape is being produced, the material can no longer be turned because the shape is a re-entrant shape, and the sections must be designed accordingly as shown in fig. 5.18. Sections such as the flanged I section shown in fig. 5.19 cannot be produced by rolling because there is a re-entrant feature in both directions.

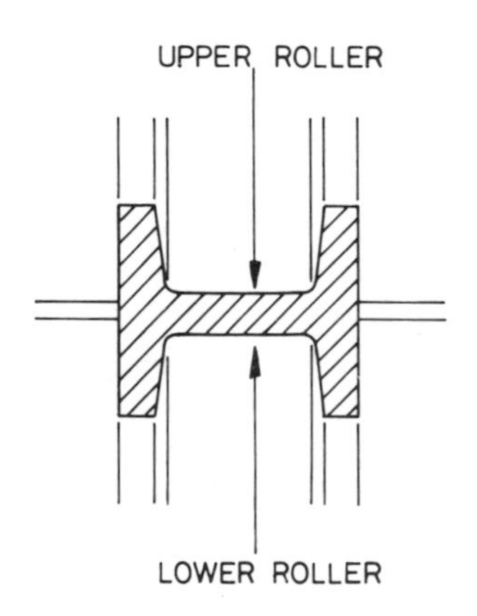

Fig. 5.18 'I' section

Rolled sections tend to be of a standard size and shape because of the cost of the rollers, the storage space required for them, and the cost of setting up the rolling mill.

5.4 Design of extruded sections

The extruding process involves pushing hot metal through a die and enables sections such as that shown in fig. 5.19 to be produced without difficulty. Extruded sections are broadly classified as solid (fig. 5.20), semi-hollow (fig. 5.21) and hollow (fig. 5.22).

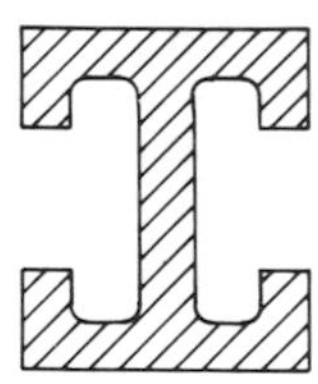

Fig. 5.19 Flanged 'I' section.

Unlike those of forging and hot-rolling, the product of extruding is dimensionally accurate and has a good quality surface finish. Extruding cannot replace hot-rolling for all products because it cannot produce sections larger than those that can be circumscribed by a circle of about 350 mm diameter. Also it is unsuitable for materials of high strength or those which cannot be made sufficiently plastic at a suitable temperature (to suit the extruding press material) and for materials that are abrasive. Steel can be extruded by coating it with glass fibre as a heat-insulator and lubricant but the process is expensive.

Fig. 5.20 Solid extruded section

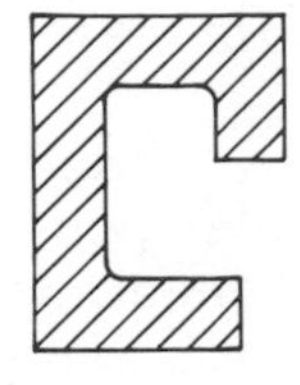

Fig. 5.21 Semi-hollow extruded section

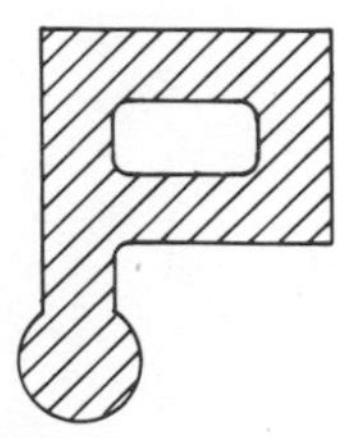

Fig. 5.22 Hollow extruded section

Fig. 5.23 Poor extruded section

Fig. 5.24 Good extruded section

Fig. 5.25 Compromise extruded section

The problem areas associated with extruding can be identified as:

(a) the behaviour of the metal during extruding;
(b) the manufacture and life of the die.

(a) The behaviour of the metal during extruding

Extruding develops a grain fibre that is similar to that developed by hot-rolling and which is therefore related to the shape of the section that is produced. To ensure uniformity across the section the material should pass through the die at the same rate at all places and therefore typical ideal sections are circular sections and rectangular sections. But extruding is usually associated with complicated shapes to combine stiffness with minimum weight and to satisfy special functional requirements (for example window frames for double glazing). Such sections should be designed to be of uniform section or of gradually changing section as shown in figs 5.23 to 5.27.

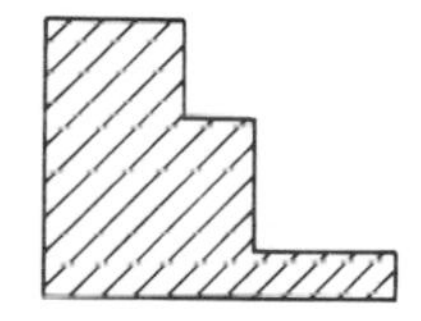

Fig. 5.26 Poor extruded section

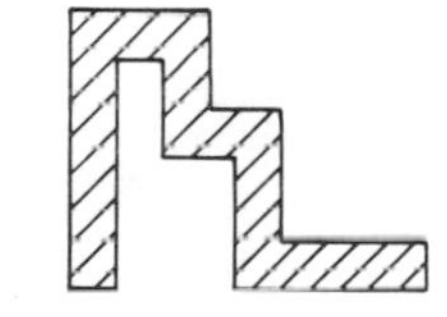

Fig. 5.27 Good extruded section

(b) Die manufacture and die life

The shape of the section should be such that the hole in the die through which the material must pass can be easily cut. With the introduction of new, non-mechanical processes for metal removal, the problem of die-manufacture has been considerably reduced. But although it is now possible to produce holes with sharp edges in the dies, sections with corner and fillet radii are preferred because both the product and the dies will be stronger. The shape should also be such that the die does not have weak portions where fracture can occur. Fig. 5.28 shows a section that requires the weak die shown in fig. 5.29, and fig. 5.30 shows an improved version that can be produced using the stronger die shown in fig. 5.31.

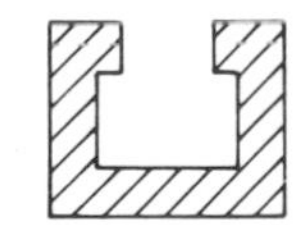

Fig. 5.28 Extruded section requiring a weak die

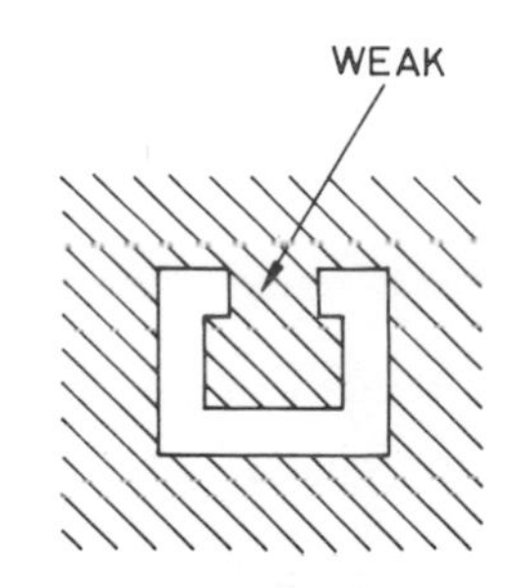

Fig. 5.29 Weak die

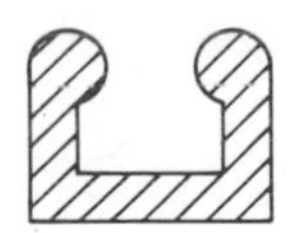

Fig. 5.30 Extruded section that does not require a weak die

Fig. 5.31 Stronger die

(c) Examples of the application extruded sections

Fig. 5.32 to 5.34 show how extruded sections are used in long lengths. Figs 5.35 to 5.37 show how extruded sections can be cut into short lengths as an alternative to using other manufacturing methods such as presswork and die-casting.

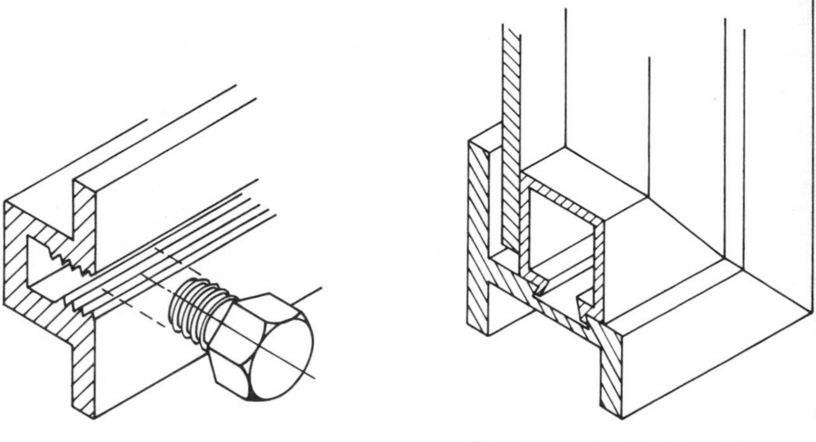

Fig. 5.32 Extruded section with screw thread form incorporated in a groove

Fig. 5.33 Extruded sections designed to snap together

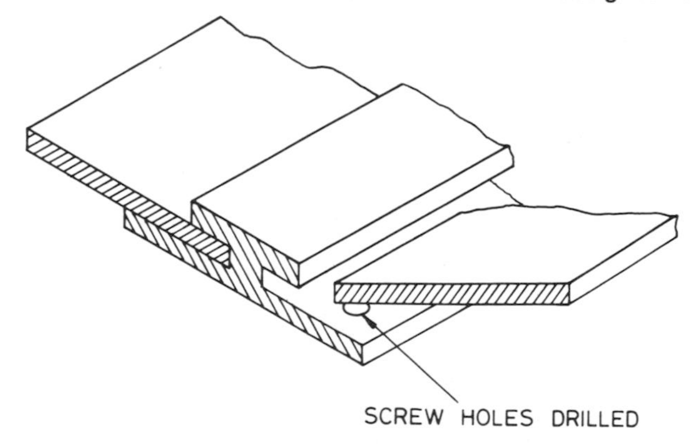

Fig. 5.34 Extruded section designed to provide invisible fixing

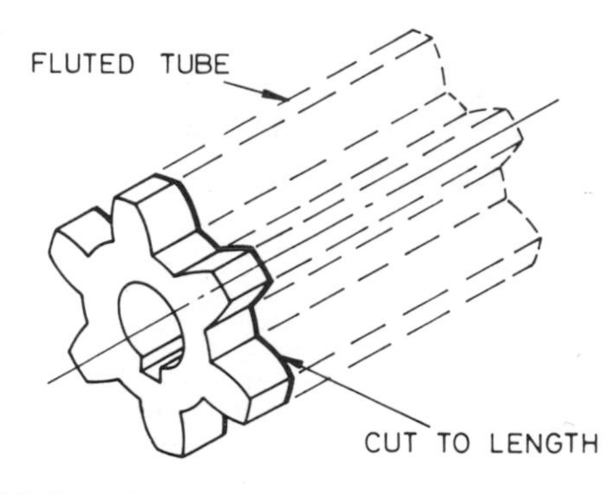

Fig. 5.35 Extruded section used to produce gears

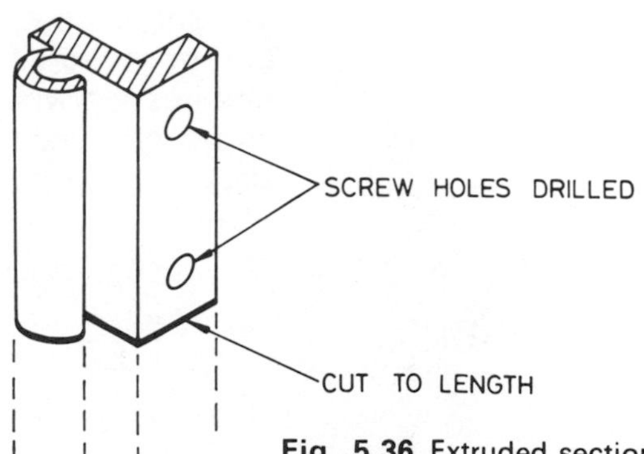

Fig. 5.36 Extruded section used to produce a door handle

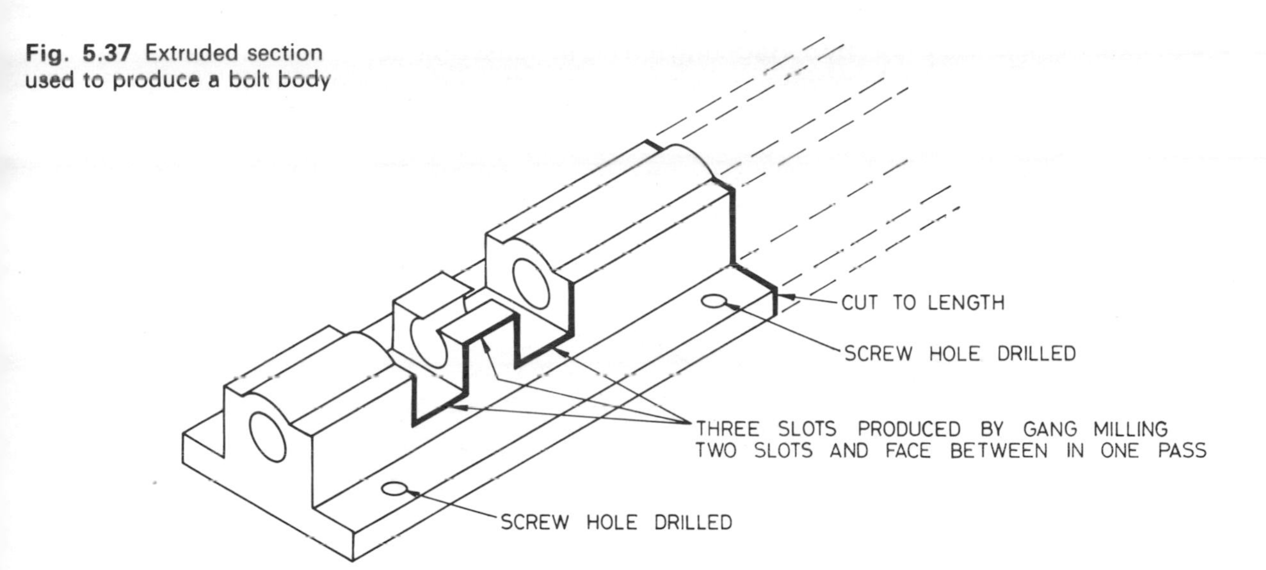

Fig. 5.37 Extruded section used to produce a bolt body

5.5 Design for cold-working

Hot-worked metal can be annealed, to make it weaker and softer, and then cold-worked. Annealing does not alter the orientation of the grains and therefore it does not remove the grain fibre produced by the hot-working. The cold-working improves the dimensional and surface accuracy of the material and also causes work-hardening.

Components are often produced from cold-rolled sheet metal using presswork. Presswork can be broadly classified into (i) cutting operations and (ii) non-cutting operations. Piercing is a cutting operation that produces holes in the workpiece and blanking is a cutting operation that cuts the workpiece from the sheet—these operations are usually combined. The principal non-cutting operations can be classified into two groups. One group consists of forming and bending; the other group consists of operations in which the metal is pushed to form a canister-like part (the operation that produces a shallow shape is called *cupping* and the operation that produces a deep shape is called *drawing*).

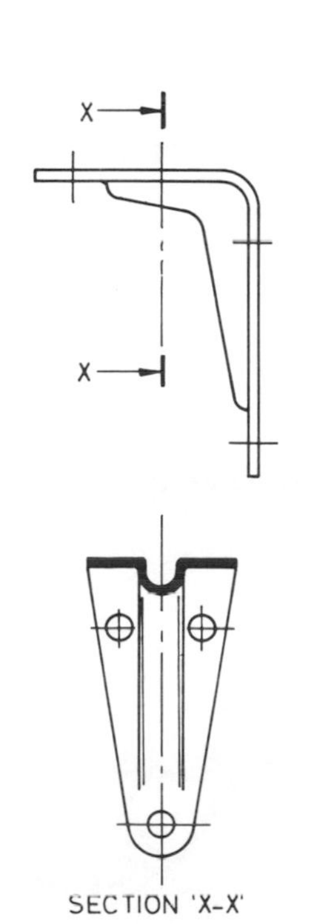

Fig. 5.38 Bracket stiffened by forming a rib

Components produced by presswork are basically of simple shape compared with those produced by operations such as forging, casting, powder metallurgy and machining. Presswork is ideal for the manufacture of tab-washers, clips, cup-like containers etc. The lack of stiffness associated with sheet metal can be overcome by the shape of the product. For example, a reasonably stiff shelf bracket can be obtained by forming it to produce a rib as shown in fig. 5.38; and a flat sheet can be stiffened by introducing a number of ribs as shown in fig. 5.39 or by corrugating it. This increases the stiffness in one direction only; stiffness in a second direction can be obtained by introducing additional ribs.

The problem areas associated with presswork can be identified as:

(a) the manufacture of the die-set;
(b) the behaviour of the metal during its manipulation;
(c) the separation of the pressed part from the die-set.

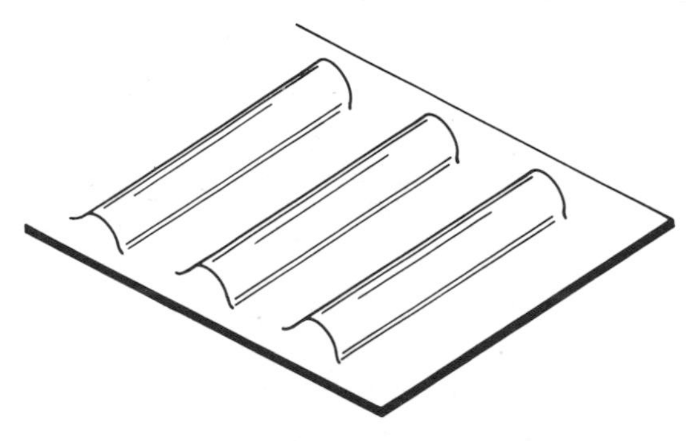

Fig. 5.39 Ribs formed to stiffen a flat sheet

(a) The manufacture of the die-set

The cost of manufacturing the die-set depends upon the complexity of the product, which should therefore be as simple as possible. Apparently minor changes can considerably reduce the cost of the die-set. For example, the rounded end shown in fig. 5.40 requires a more costly die-set than does the square end shown in fig. 5.41; the 45° chamfers shown in fig. 5.42 are a compromise. When rounded ends are essential it must be remembered that the tangential radii shown in fig. 5.40 cost more than do those shown in fig. 5.43. The tooling costs can be reduced by designing the product so that it can be produced using the minimum number of forming operations as illustrated by figs 5.44 and 5.45.

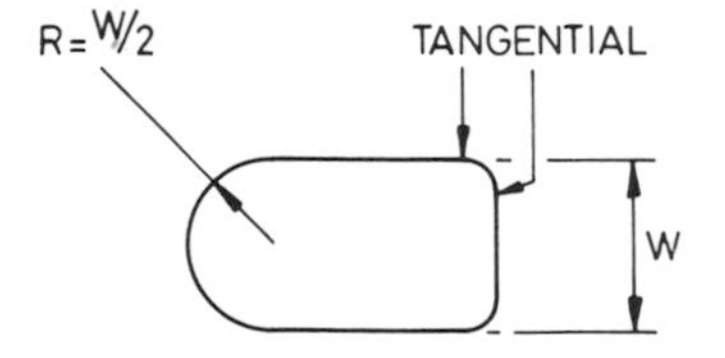

Fig. 5.40 Unsuitable profile

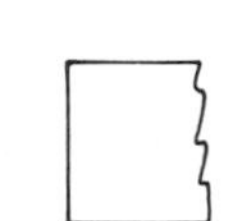

Fig. 5.41 Suitable profile

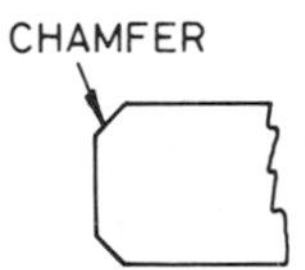

Fig. 5.42 Compromise profile

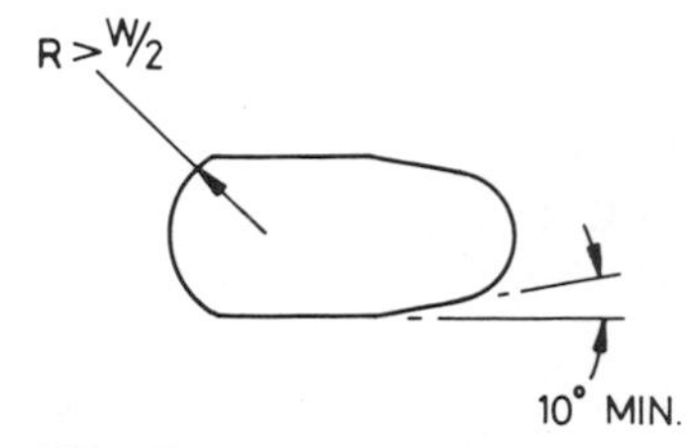

Fig. 5.43 Compromise profile with radii

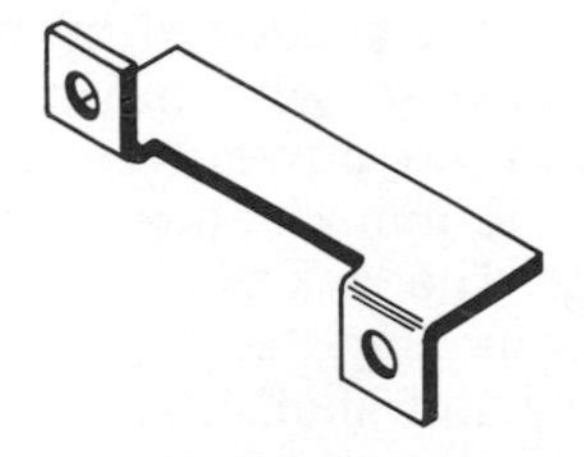

Fig. 5.44 Pressing with two-directional forming

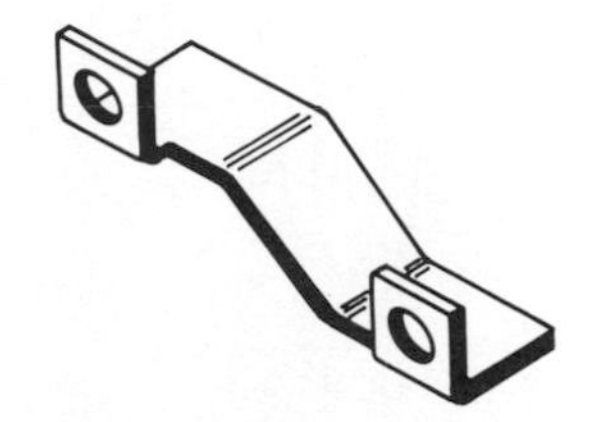

Fig. 5.45 Pressing with single direction forming

(b) The behaviour of the metal during its manipulation

Annealing and cold-working do not remove the directionality produced by previous hot-working—in fact the cold-rolling done to prepare the sheet for presswork increases the directional effect by deforming the grains. The directional properties should be taken into account when designing parts that are produced by bending. Consider the component shown in fig. 5.46. If the material is blanked prior to bending using the layout shown by A in fig. 5.47 there will be no trouble; but if blanked using the layout shown by B in fig. 5.47 it will probably split when bent because the line of the bend lies in the same direction as the grain fibre. In this example the problem is overcome by using a suitable blanking layout.

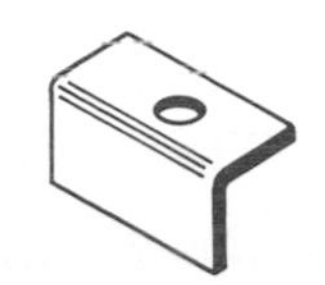

Fig. 5.46 Sheet-metal component

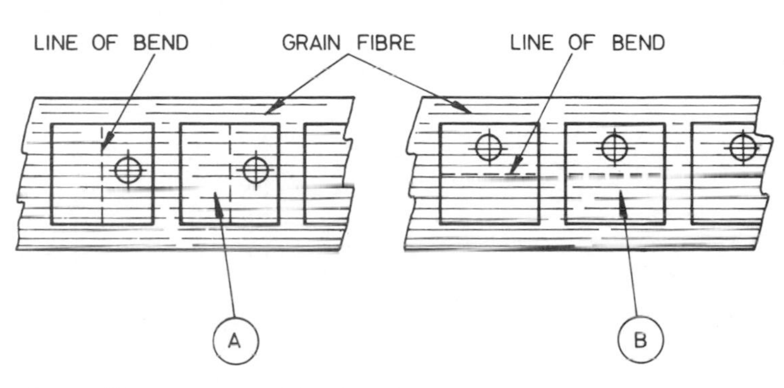

Fig. 5.47 Piercing and blanking layout

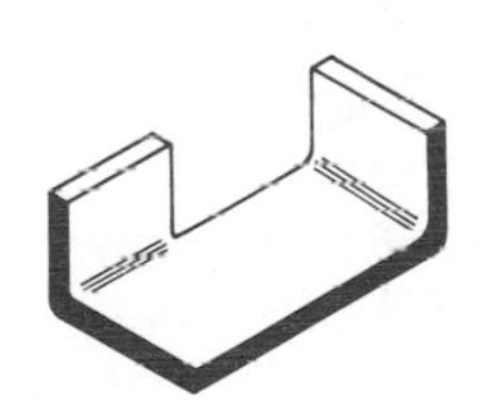

Fig. 5.48 Pressing that would be troublesome to produce

A component such as that shown in fig. 5.48 causes problems because it has two tabs, one of which must be in an unfavourable position; but by changing the design to that shown in fig. 5.49 the problem is eliminated.

Abrupt bends should be avoided because they tend to produce stress concentration. Similarly the metal will tear when bent to form the tab shown in fig. 5.50. The tearing will not occur if relief notches shown in fig. 5.51 are introduced or if the form is modified to that shown in fig. 5.52.

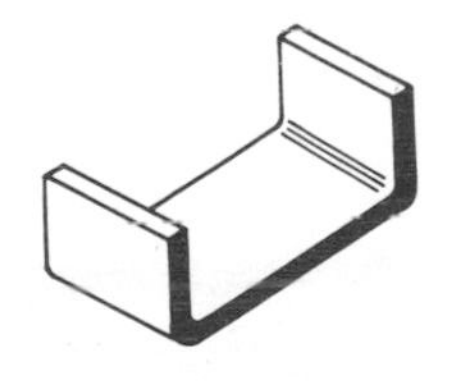

Fig. 5.49 Re-designed pressing

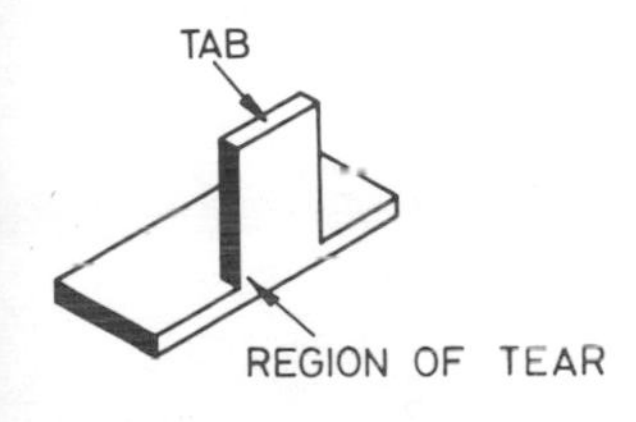

Fig. 5.50 Pressing that would tear

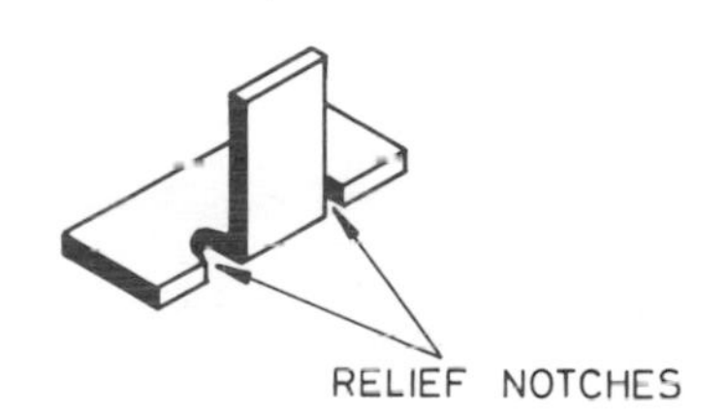

Fig. 5.51 Re-designed pressing

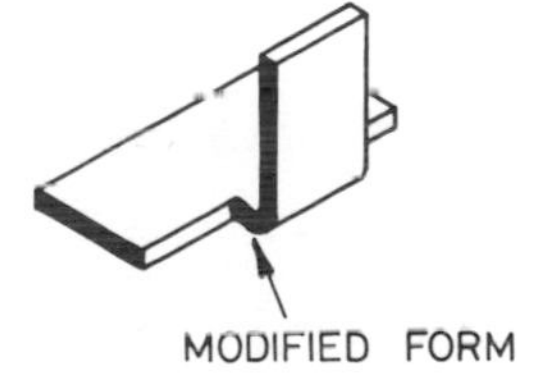

Fig. 5.52 Re-designed pressing

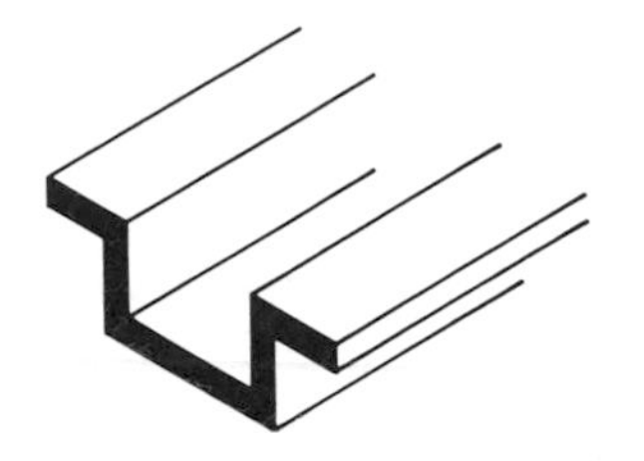

Fig. 5.53 Pressing that would be difficult to remove from the die-set

(c) The separation of the pressed part from the die-set

The pressing should be designed so that it can be easily removed from die-set. For example, the pressing shown in fig. 5.53 is more difficult to remove from the die-set than that shown in fig. 5.54. Similarly the pressing shown in fig. 5.55 can be produced by forming the two flanges with slide tools, but it must be slid from the punch by the operator. The modified version shown in fig. 5.56 has two external flanges that can be formed by bending before the U bend is done (a double-acting press would be used) and the completed pressing removed from the punch during its upward movement by a stripper plate. A springy pressing, such as a fastener, may be of a re-entrant shape if it can be sprung open by a stripper plate to release it from the punch.

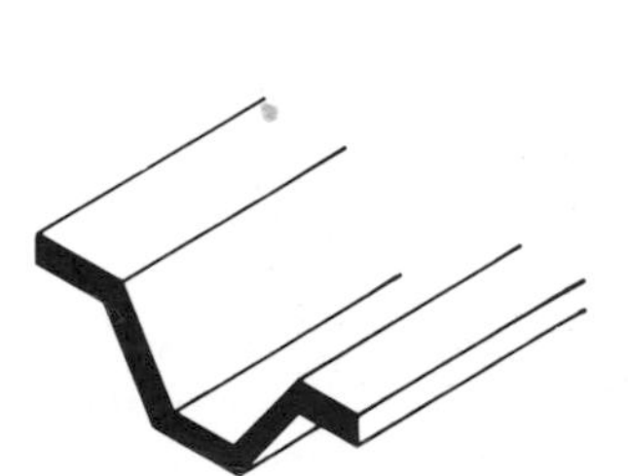

Fig. 5.54 Re-designed pressing

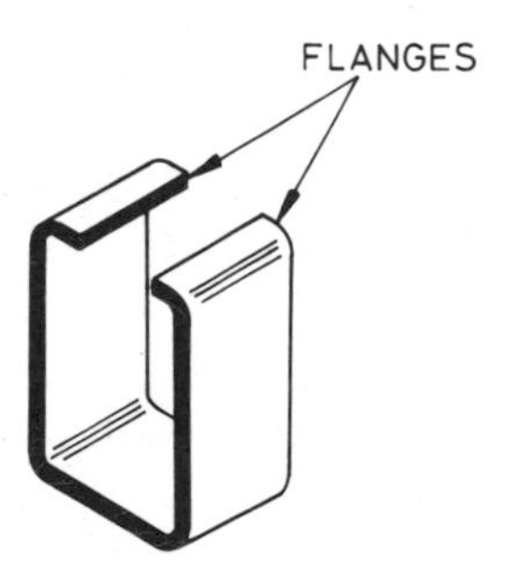

Fig. 5.55 Awkward pressing

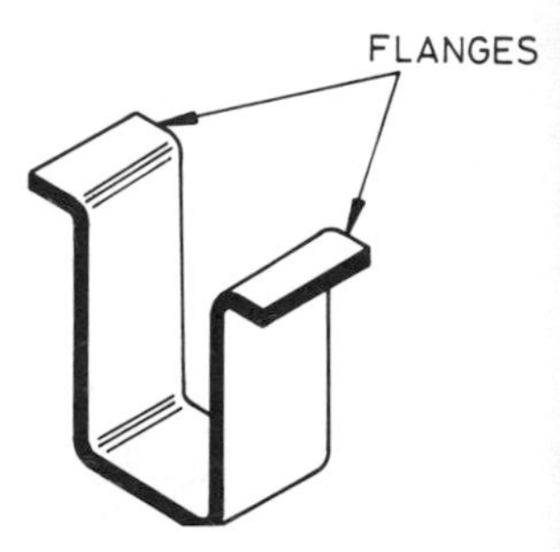

Fig. 5.56 Re-designed pressing

6 Design for Powder Metallurgy

Graded and blended metal in powder form can be manipulated into shape by compacting it, to cold-weld the particles together, and then sintering the compact at a temperature of about 75% of the melting point of the metal.

The powder is usually compacted in a vertical press. It is dispensed into a die and pressure exerted on it by two punches to obtain a shape that is of the correct size and weight to produce a finished part with the specified density. The compact so produced is ejected from the die by the lower punch which moves upwards.

Powder metallurgy is a rapid method of producing fairly small, accurate parts in most metals (aluminium and its alloys being notable exceptions). Powder metallurgy can give a mixture of metals that will not alloy in the conventional way and produce a conglomerate of metal and metallic compounds (for example the conglomerate of tungsten carbide and cobalt used as a cutting tool material). Also the degree of porosity can be controlled (for example when a filter or a porous plain bearing is produced). Parts produced by powder metallurgy are weak compared with those produced by other methods. Also the cost of tooling required to compact the powder is high and so the process is only economical when large quantities are produced.

The problem areas associated with powder metallurgy can be identified as:

(a) the behaviour of the powder during the compacting operation;

(b) the ejection of the compact from the die;

(c) the manufacture and working life of the die set.

These problems are not considered chronologically because the design priority order is not chronological. In addition to satisfying the basic problems, the design must be such that full advantage is taken of the ability of the process to produce complicated shapes, intricate detail and certain shapes that cannot be obtained by machining.

6.1 Behaviour of the powder during compacting

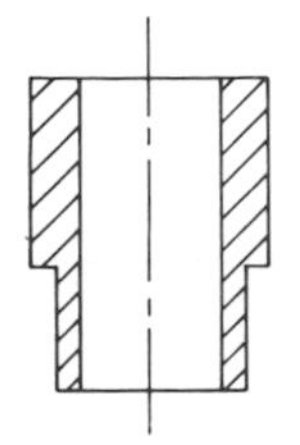

Fig. 6.1 Unacceptable design because of local thin section

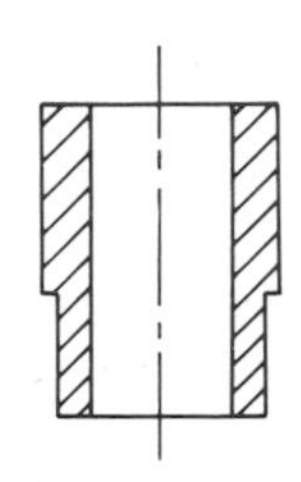

Fig. 6.2 Acceptable design

Unlike liquid metal, which flows freely during casting, or even solid metal, which can be made to flow when it is wrought, powder metal does not flow. This means that thin walls are difficult to produce, that variations in section thickness will result in non-uniformity of structure, and that stepped diameters are only acceptable if the variation in diameter is not too great, or the section is not too thin (see figs 6.1 and 6.2).

The structure will be non-uniform if the ratio of length, in the direction of pressing, to diameter exceeds about $2\frac{1}{2}:1$ because of the variation of density this will cause. (This ratio may be increased to 4:1 without the variation causing trouble if the diameter is large.)

Holes that lie in the direction of pressing can be produced by introducing core rods in the die in the direction of pressing. These restrict the space into which the powder can fall when dispensed into the die. Holes so produced cannot be of very small diameter because they would require thin, and therefore fragile, core rods to produce them.

Features such as blind holes can be formed either by the top punch or by the bottom punch. When formed by the top punch the powder is moulded and the features must be shallower and of more restricted shape than when formed by the lower punch. The lower punch produces such features not by moulding the powder but mainly by controlling the space into which the powder can fall. The location and the shape of these features is closely related to the ejection of the compact from the die.

6.2 Ejection of the compact from the die

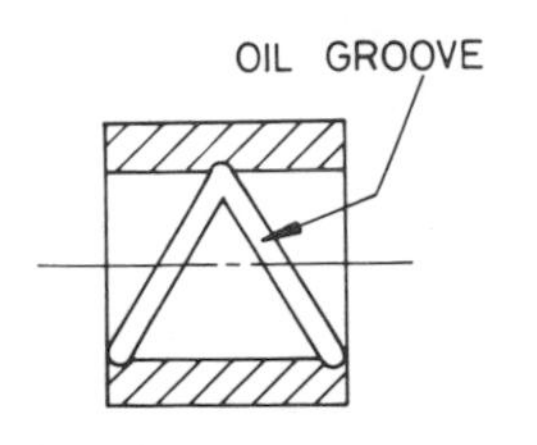

Fig. 6.3 Unacceptable design because the oil groove is a re-entrant feature

In order to allow the compact to be ejected from the die by the lower punch during its upward stroke it must have no re-entrant features. In this respect the problems associated with this process are similar to those associated with pressure die-casting. Figs 6.3 and 6.5 illustrate some re-entrant features and figs 6.4, 6.6 and 6.7 show how the designs may be improved to eliminate these features. Certain features, such as screw threads, knurls and undercuts, and those shown in fig. 6.8 can, if necessary, be produced by machining the part after it has been sintered.

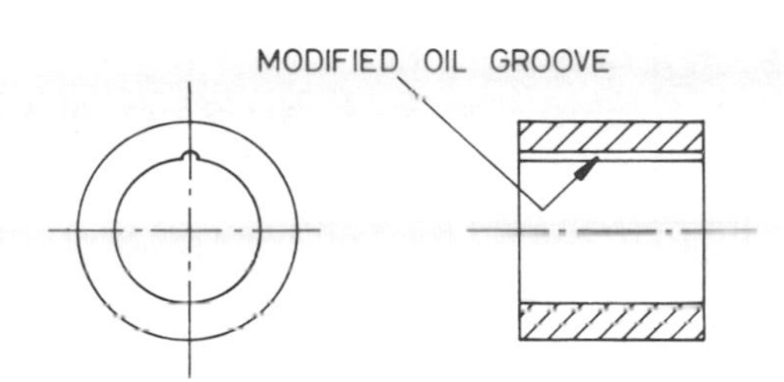

Fig. 6.4 Acceptable form of oil groove

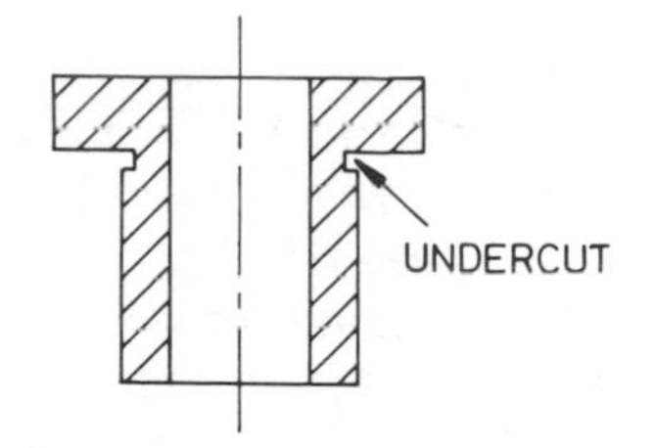

Fig. 6.5 Unacceptable design because the undercut is a re-entrant feature

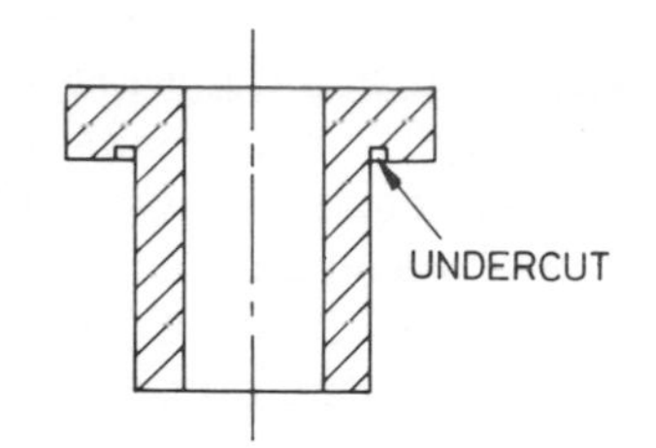

Fig. 6.6 Acceptable form of undercut

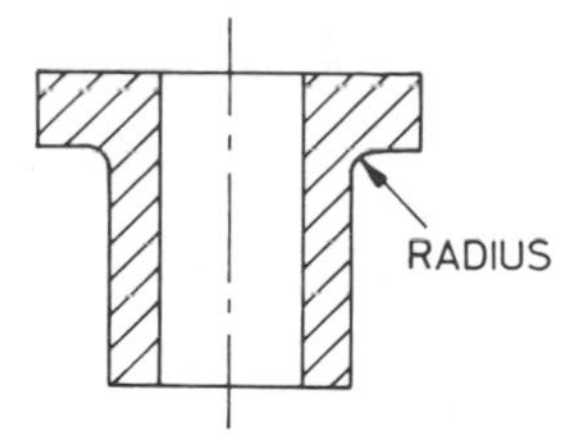

Fig. 6.7 Acceptable design with radius under the head

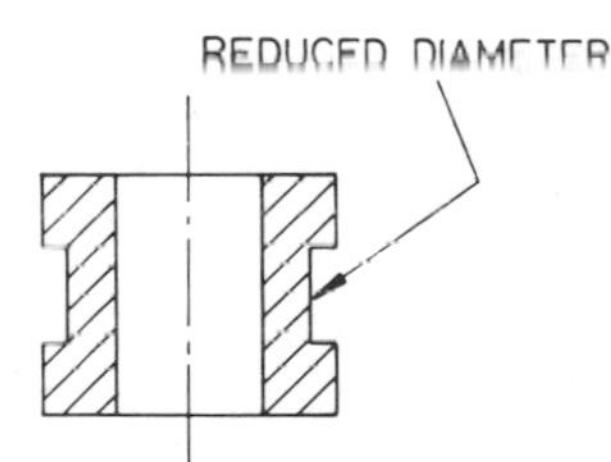

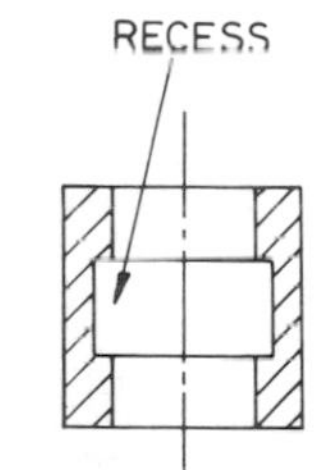

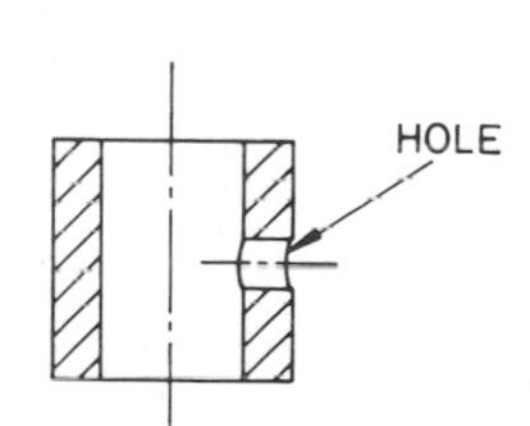

Fig. 6.8 Typical features that can be produced by machining after sintering

A part that has a re-entrant feature and a deep recess or blind hole causes design problems. For example, the part shown in fig. 6.9 cannot be compacted because the flange must be formed at the top of the die to prevent a re-entrant effect, and the blind hole is too deep to be produced by the top punch. Fig. 6.10 shows the modified design that must be used if both the flange and the blind hole are to be retained; fig. 6.11 shows the shallow recess that must replace the blind hole if the recess and flange must be at the same end.

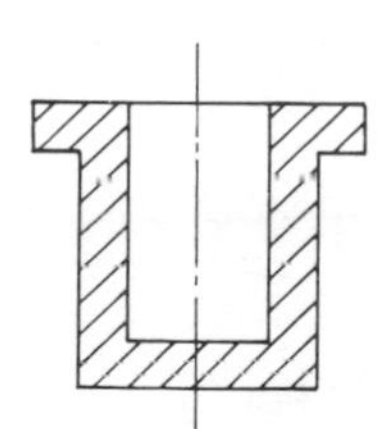

Fig. 6.9 Unacceptable design

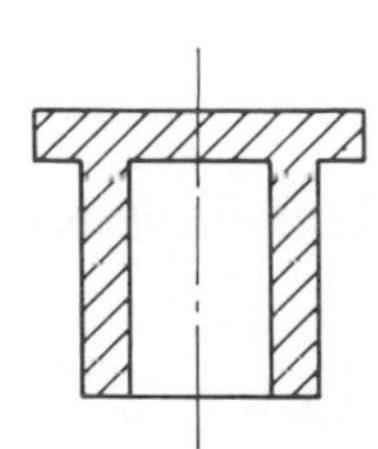

Fig. 6.10 Acceptable design

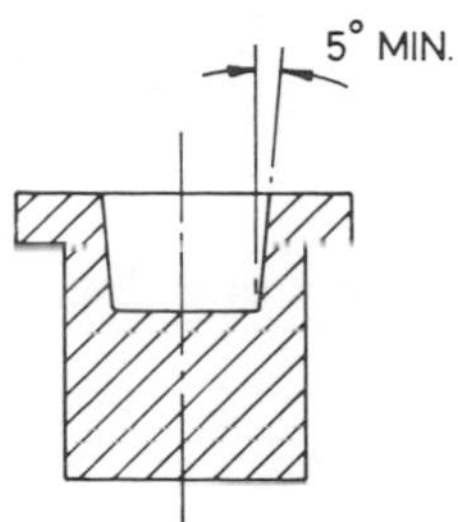

Fig. 6.11 Acceptable design

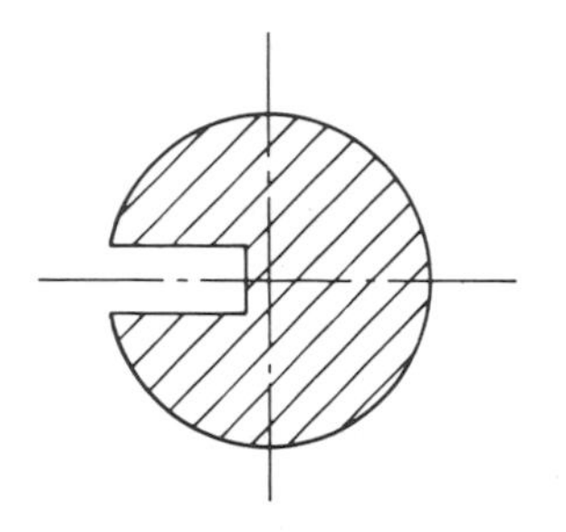

Fig. 6.12 Unsuitable design because it requires a die that is weak

6.3 Design to simplify the manufacture of the die-set and to extend its working life

The part should be designed so that the punches and the die do not have thin sections or knife edges. These would make the manufacture costly and reduce the working life of the die set because of inaccuracy as a result of wear.

For example the deep narrow slot in the part shown in fig. 6.12 would require the die to have a weak projecting feature to produce it, but the shallow wide slot in fig. 6.13 can be produced by a much stronger die. Die sets that are used to produce profiles with feather edges, as illustrated in fig. 6.14, will have local weakness, but by a small change of profile, as shown in fig. 6.15, the local weakness can be eliminated. The spherical part shown in fig. 6.16 can only be compacted using a die set that includes

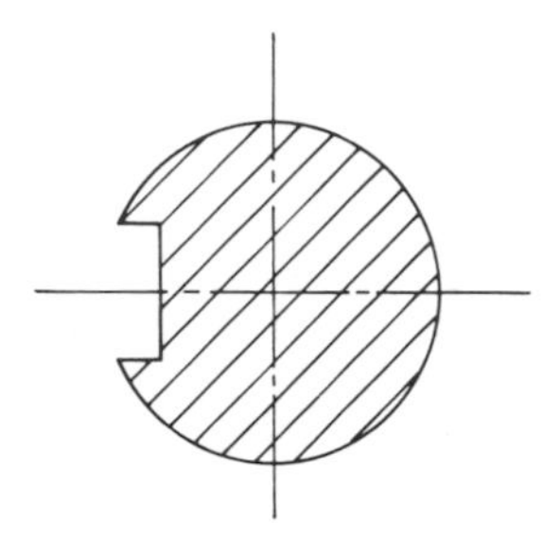

Fig. 6.13 Suitable design using a stronger die

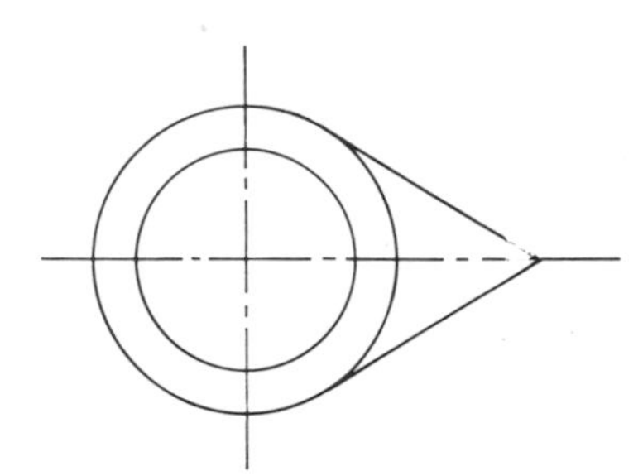

Fig. 6.14 Unsuitable design requiring a die set with knife edges

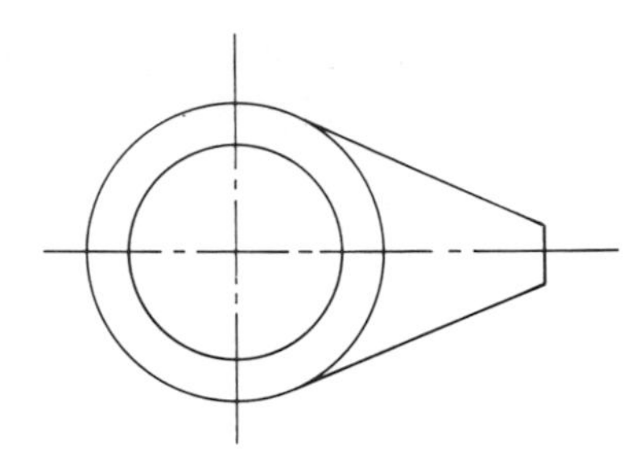

Fig. 6.15 Suitable design

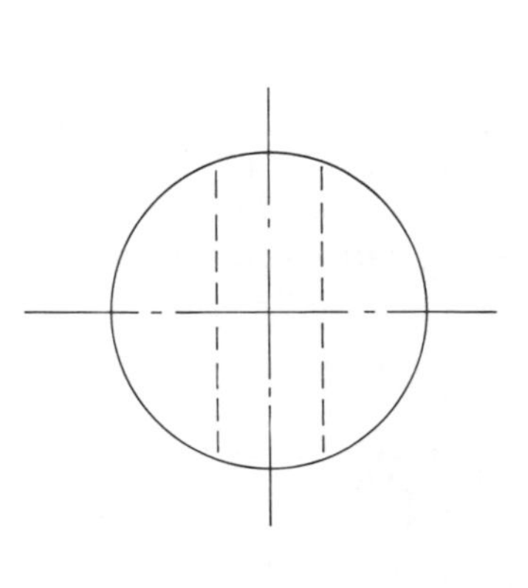

Fig. 6.16 Spherical part

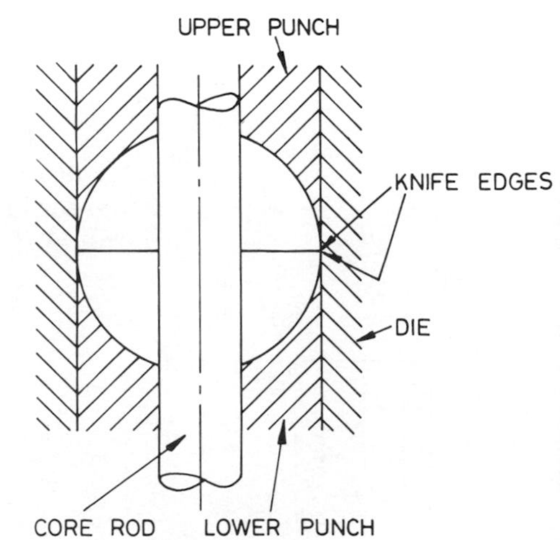

Fig. 6.17 Die set to produce a spherical part

punches with knife edges as shown in fig. 6.17. These would be easily damaged by collision at the end of the stroke. By introducing a cylindrical portion, as shown in fig. 6.18, the knife edge effect is reduced, as shown in fig. 6.19, and the collision problem eliminated. The design can be further improved by including a raised portion, illustrated in fig. 6.20, that can be removed by a post-sintering operation such as sizing, and which enables the punch to have a land instead of a sharp edge.

Countersunk holes that are produced by compacting should be different from those produced by machining. The countersunk hole in fig. 6.21 could be easily produced by machining but requires a punch with a knife edge as illustrated in fig. 6.22. The countersunk hole in fig. 6.23 would be unsuitable for production by machining but does not require the use of a punch with a knife edge (see fig. 6.24).

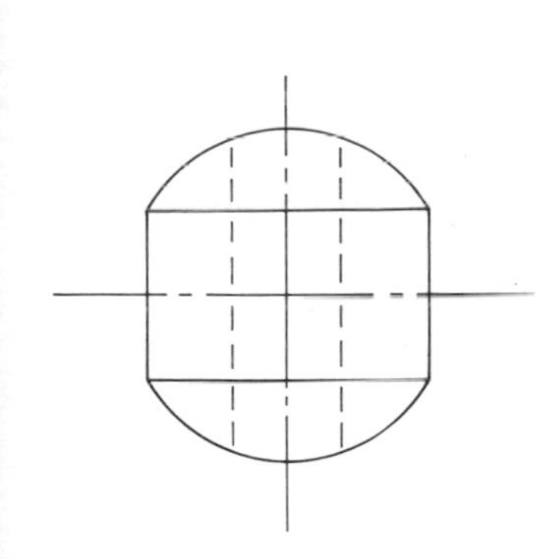

Fig. 6.18 Modified spherical part

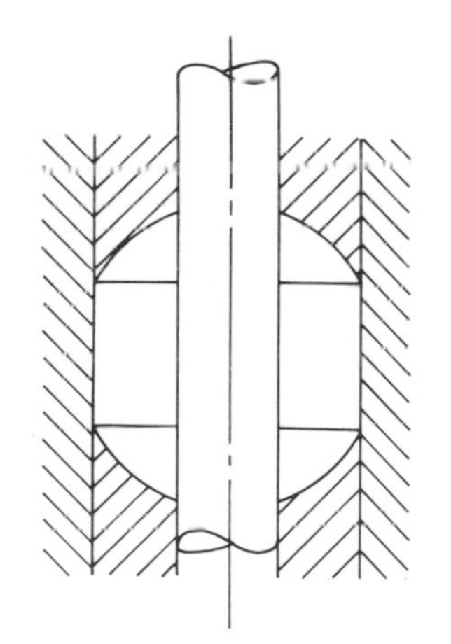

Fig. 6.19 Die-set to produce modified spherical part

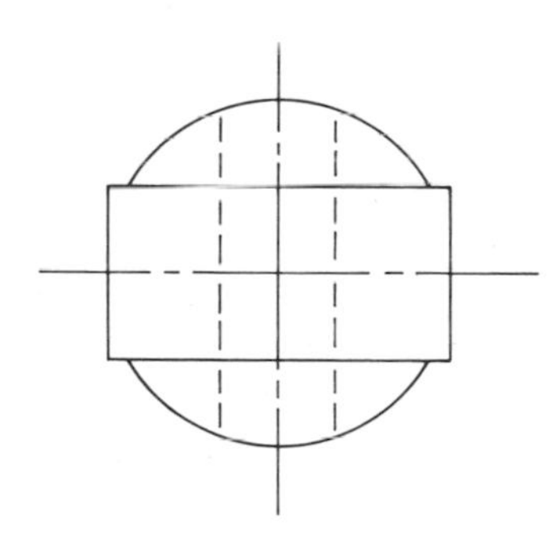

Fig. 6.20 Further-improved part

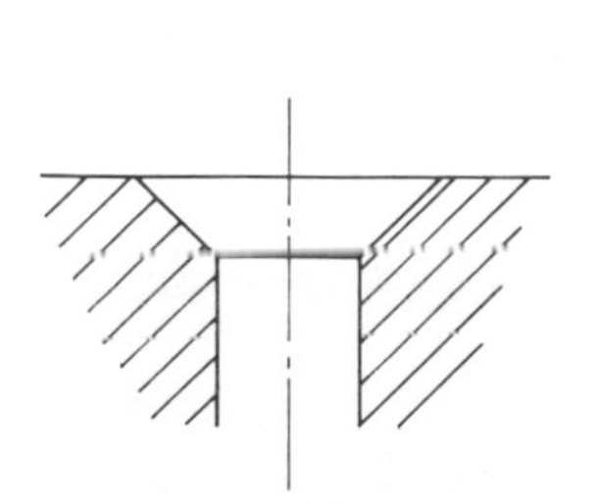

Fig. 6.21 Countersunk hole to suit machining

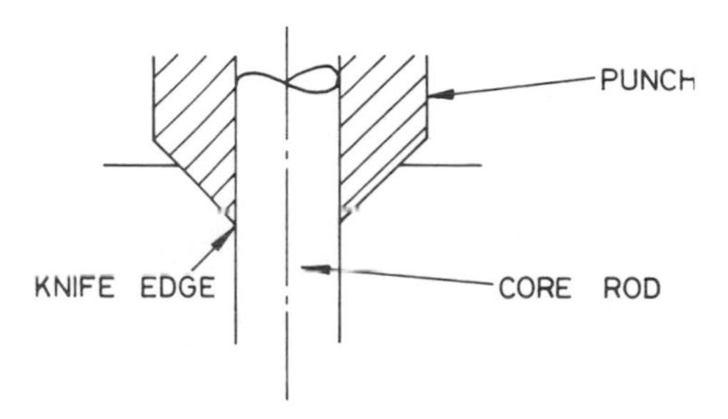

Fig. 6.22 Punch to suit countersunk hole

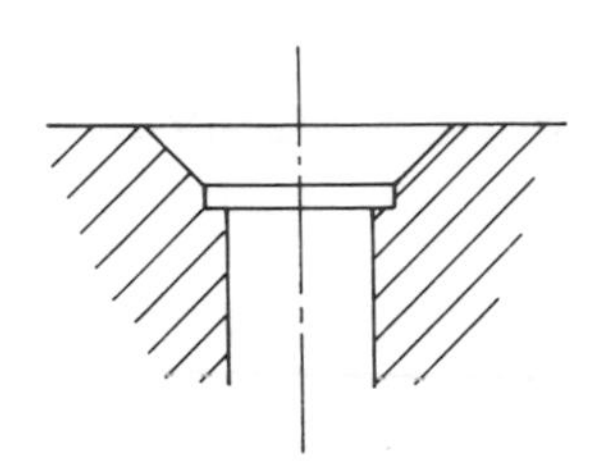

Fig. 6.23 Modified countersunk hole

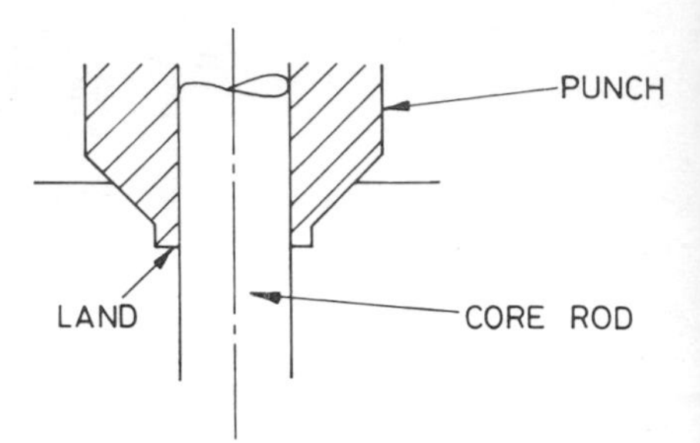

Fig. 6.24 Punch to suit modified countersunk hole

6.4 Accuracy

The accuracy of parts produced by powder metallurgy is better at right angles to the direction of pressing than in the direction of pressing. This is because the former is controlled almost completely by the accuracy of the press tool set, but the accuracy in the direction of pressing depends, to a large extent, upon that of press setting. The accuracy of a sintered part can be improved by pressing it between dies (coining) in the direction of pressing at the compacting operation, or by passing it through a die (sizing). The following tolerances are those of products before these additional operations are done.

- Tolerances in the direction of pressing ±0.1 mm per 25 mm.
- Tolerances at right angles to the direction of pressing ±0.01 mm per 25 mm.
- Eccentricity of circular features 0.08 mm.

7 Design for Machining

Machining is used to produce shapes and a degree of accuracy that cannot be directly obtained by primary manipulation methods such as forging, casting and powder metallurgy. It is also used to produce parts by machining from billets or from bar stock. The problems identified in this chapter are those associated with mechanical metal removal (for example, turning, milling, drilling and grinding). Some of these problems are overcome by the newer, more costly and less commonly used thermal and chemical methods of metal removal.

There are three problem areas that affect the shape of the product.

(a) The limitations of shape imposed by the need to relate the product to the relative workpiece and cutting tool movements provided by the machine, the accuracy that can be obtained and the cost of the operation.

(b) The limitations of shape imposed by the need to locate, seat and clamp the workpiece during machining.

(c) The limitations of shape imposed by the need to provide the cutting tool with space for approach and over-run.

In addition to allowing for these limitations, the design must be such that the cost of machining is minimised. This can be achieved by keeping the shape as simple as possible and by keeping down the amount of metal that must be removed and the number of machine settings that are required.

7.1 Design to suit the relative workpiece and cutting tool movements

The shape to be produced by machining must be such that it can be obtained by the workpiece and cutting tool movements of the machine (or machines) that are available. For example, the basic movements of a lathe are rotation of the workpiece and linear movement of the cutting tool. This movement may be parallel to the axis of rotation or at right angles to it. Other cutting tool movements can be obtained by setting the machine slides or by using a copying system. Shapes that can be directly produced by the basic machine movements are the most convenient and those requiring the use of a copying system are the least convenient. The order of preference regarding shape to suit lathework is as follows.

(i) First choice. The first choice (see fig. 7.1) is a shape that consists of concentric cylindrical surfaces (D_1, D_2, and D_3),

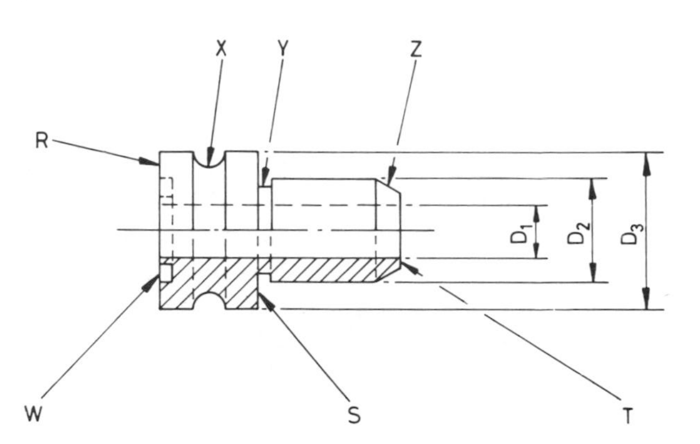

Fig. 7.1 First choice of shape for lathework

surfaces that are at right angles to the axis (R, S, and T), and recesses, chamfers etc. (W, X, Y, and Z) that can be produced using form tools and the basic movements. Screw threads are a special case of cylindrical shape.

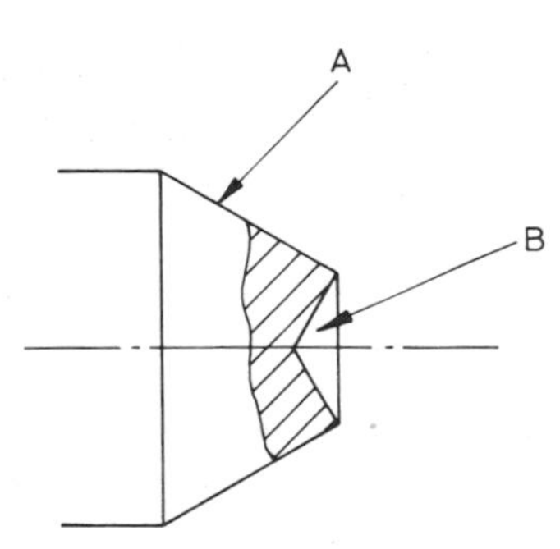

Fig. 7.2 Second choice of shape for lathework

(ii) Second choice. This includes concentric conical surfaces (A and B of fig. 7.2) that are too long to be produced using a form tool because of chatter. These require the setting of the compound slide, the use of a taper turning attachment, or the setting over of the tailstock, to obtain a cutting tool movement that is inclined to the axis.

(iii) Third choice. Cylindrical features that are eccentric with respect to the main axis (see fig. 7.3) require the setting over of the workpiece so that the axis of the eccentric feature(s) coincides with that of the machine (see dimension E). This requires an additional machining operation, or operations, and either individual setting or the use of a fixture. The problem of the workpiece being out of balance during machining must also be considered.

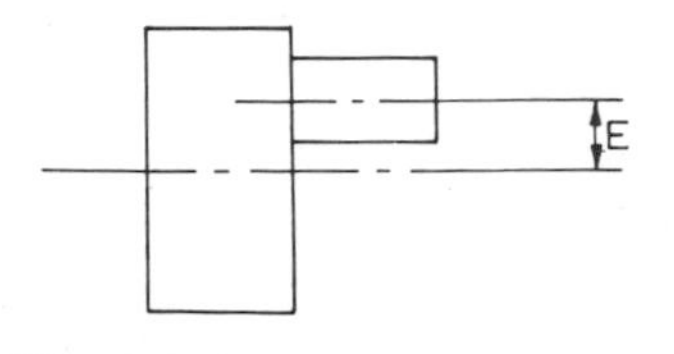

Fig. 7.3 Third choice of shape for lathework

(iv) Fourth choice. Surfaces that involve a frequent or continuous change of cutting tool direction usually cannot be obtained

by the basic machine movements or by changes in setting. Form tools can be used if the surface is short, or if a long contour is broken up by recesses to enable short-contact form tools to be used without the inevitable step between each section of the contour. A long surface as shown in fig. 7.4 can only be produced by one cutting tool with its movement controlled by a copying system.

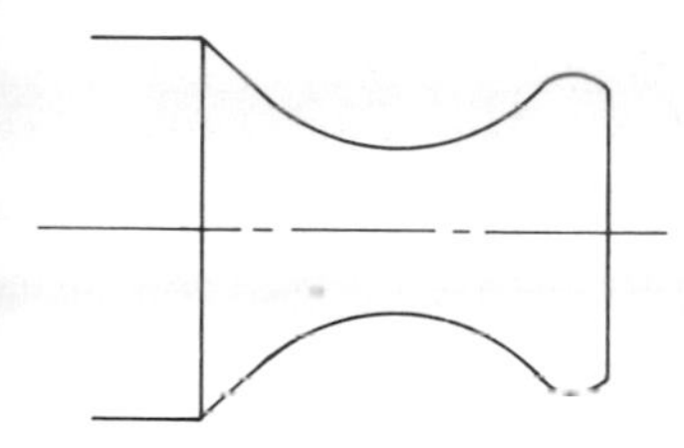

Fig. 7.4 Fourth choice of shape for lathework

It will be appreciated that the function of a component may demand a shape that is less convenient to produce, but very often the problem can be minimised by careful consideration at the initial design stages.

7.2 Design to suit location, seating and clamping

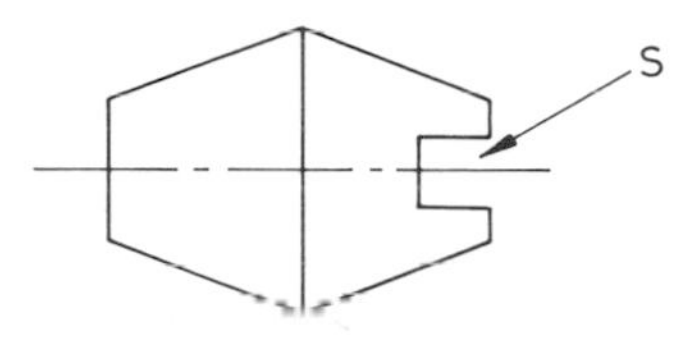

Fig. 7.5 Double-cone component

The basic requirements that must be met before a workpiece is machined are that it is located with respect to the path of the cutting tool (or its own path with respect to the cutting tool), that it is seated so that it does not deflect under the cutting forces and that it is clamped with sufficient force to prevent it from moving during machining.

Ideally a location point, or points, should be established at the first machining operation and be used at all subsequent operations. If a suitable location feature is not part of the basic design of a part it may be necessary to introduce one to assist in machining. Similarly, although a part may not have to be seated upon assembly it may be necessary to introduce a suitable feature for seating and clamping at the machining operations.

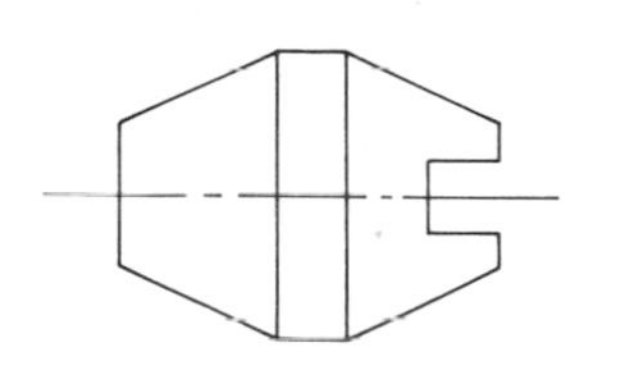

Fig. 7.6 Modified design to facilitate location and clamping

For example, fig. 7.5 shows a double-cone component with a slot S that is to be milled in a later operation. At the milling operation it could be located from one cone (although variation of its diameter will cause axial variation) and clamped from the other. The design shown in fig. 7.6 includes a cylindrical feature that can be used for location and clamping, and that in fig. 7.7 includes a cylindrical hole that can be used for location and a flange that can be used for seating and clamping.

7.3 Design to suit cutting tool approach and over-run, reduce deflection and cut costs

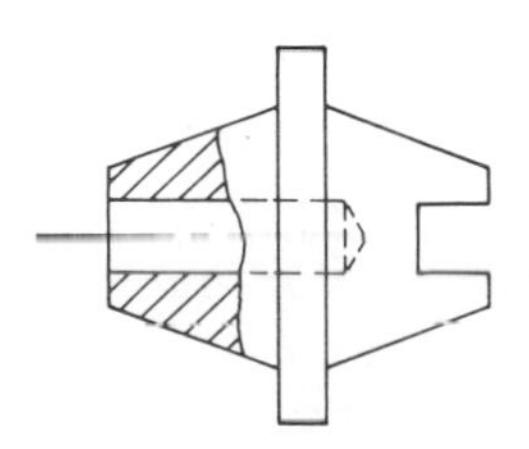

Fig. 7.7 Improved design to facilitate location, seating and clamping

(a) Design to suit cutting tool approach

Fig. 7.8 shows part of a component that includes a large conical hole. The hole must be produced by a cutting tool approach as shown, but the component also has a shallow cylindrical recess that obstructs the tool approach. The version shown in fig. 7.9 allows the tool to approach, and that shown in fig. 7.10 combines unobstructed tool approach with a shallow recess.

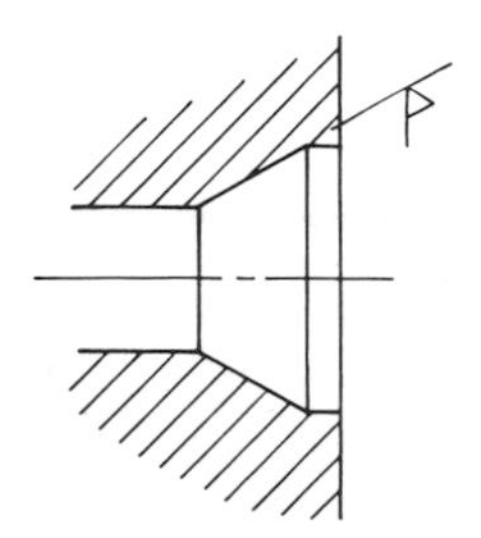

Fig. 7.8 Design that obstructs tool approach

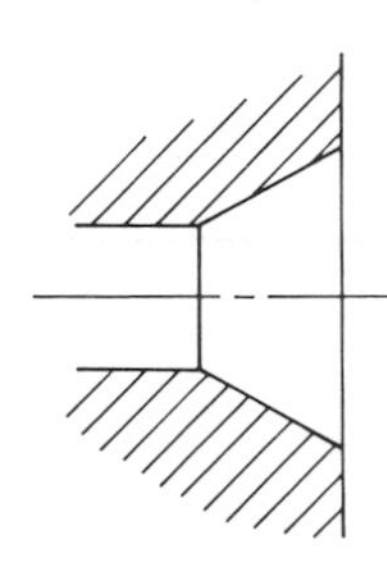

Fig. 7.9 Design that eliminates obstruction of tool approach

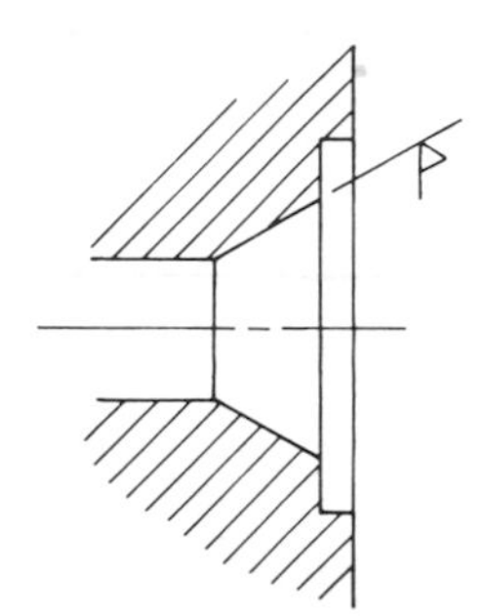

Fig. 7.10 Modified design that includes the recess

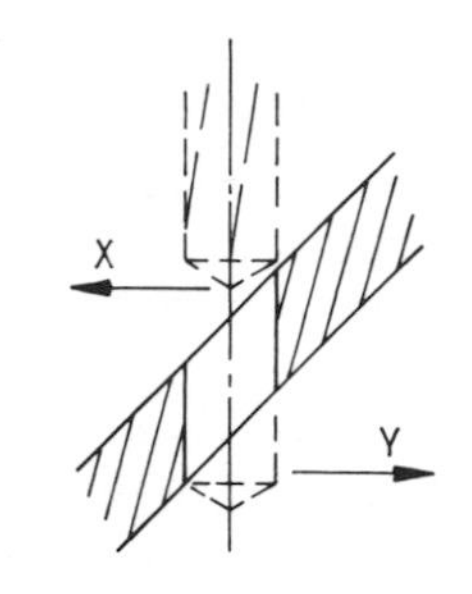

Fig. 7.11 Drilling into an inclined wall

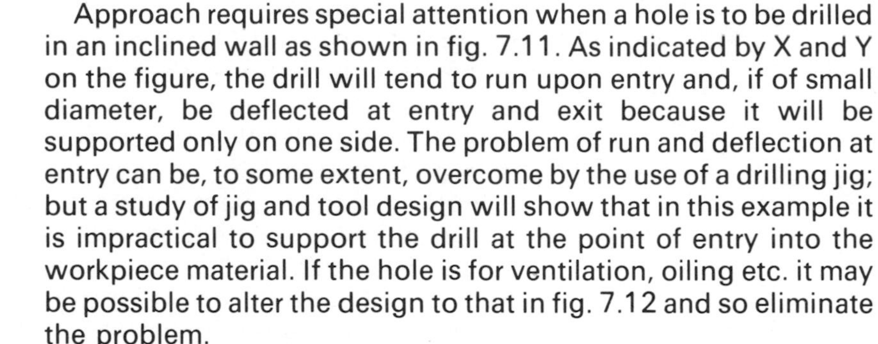

Approach requires special attention when a hole is to be drilled in an inclined wall as shown in fig. 7.11. As indicated by X and Y on the figure, the drill will tend to run upon entry and, if of small diameter, be deflected at entry and exit because it will be supported only on one side. The problem of run and deflection at entry can be, to some extent, overcome by the use of a drilling jig; but a study of jig and tool design will show that in this example it is impractical to support the drill at the point of entry into the workpiece material. If the hole is for ventilation, oiling etc. it may be possible to alter the design to that in fig. 7.12 and so eliminate the problem.

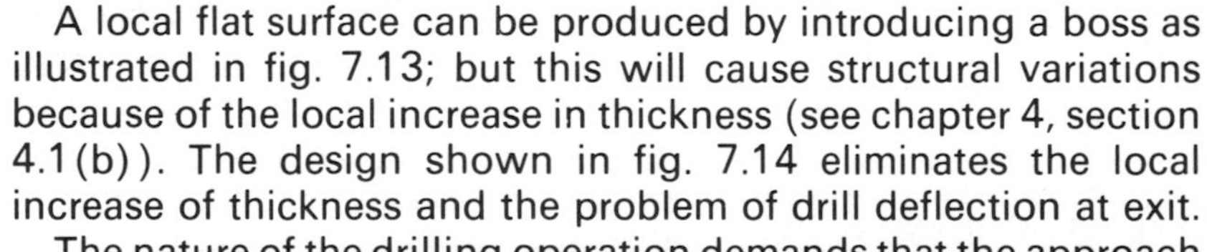

A local flat surface can be produced by introducing a boss as illustrated in fig. 7.13; but this will cause structural variations because of the local increase in thickness (see chapter 4, section 4.1(b)). The design shown in fig. 7.14 eliminates the local increase of thickness and the problem of drill deflection at exit.

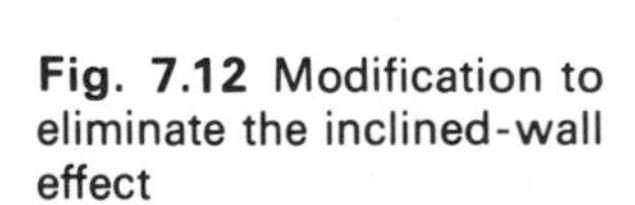

Fig. 7.12 Modification to eliminate the inclined-wall effect

The nature of the drilling operation demands that the approach path of the drill must not be obstructed. Fig. 7.15 illustrates a design in which the opening at the top of the casting is too small to allow the drill to enter without drilling into the surrounding

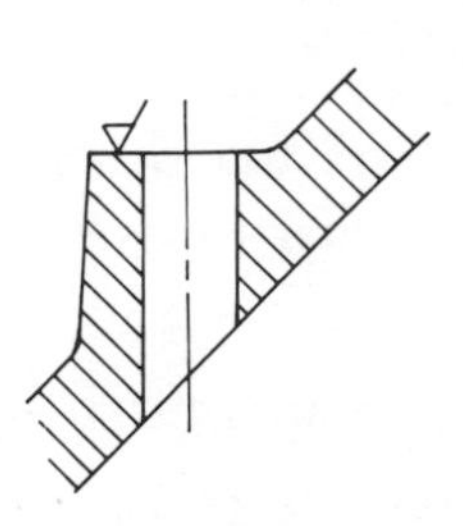

Fig. 7.13 Boss to produce a flat surface

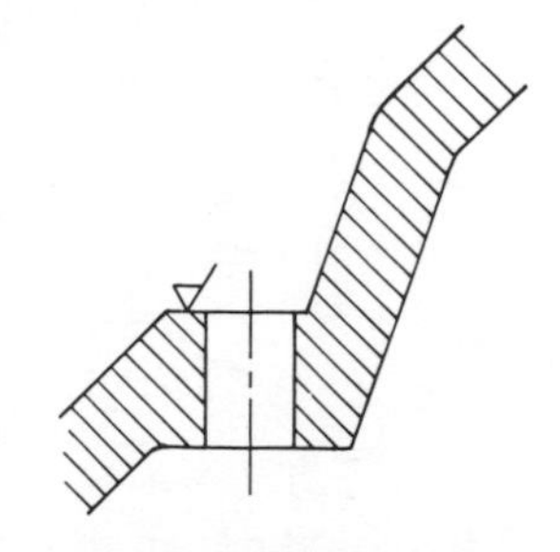

Fig. 7.14 Modified form to produce a flat surface

material, and it is impractical to approach from the other side because of the 'sloping face' effect already discussed. The problem can be solved by either enlarging the opening, as shown in fig. 7.16, or by re-designing the wall, as shown in fig. 7.17, to enable the drill to enter from the opposite side.

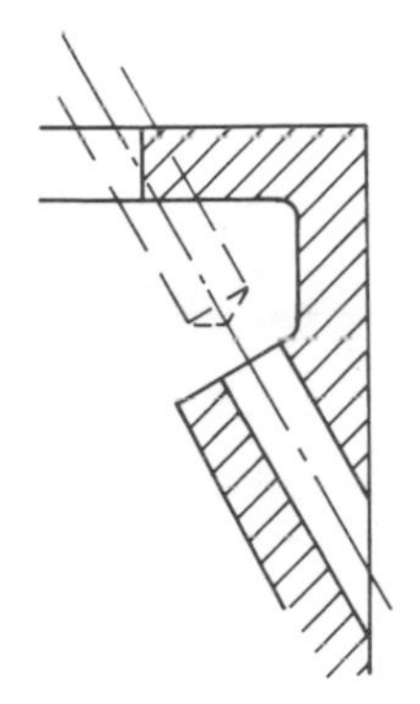

Fig. 7.15 Design with obstructed drill path

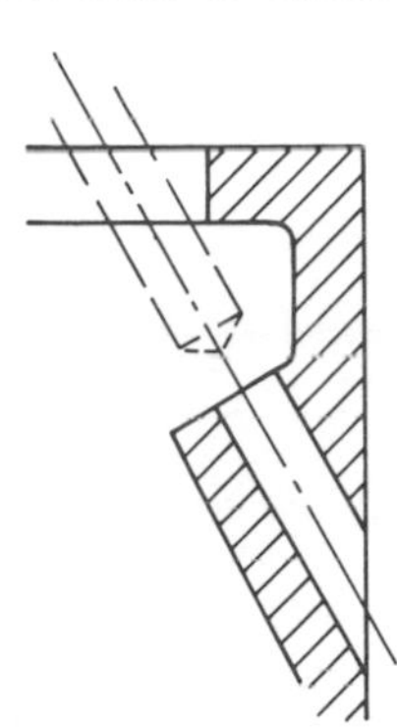

Fig. 7.16 Design with unobstructed drill path

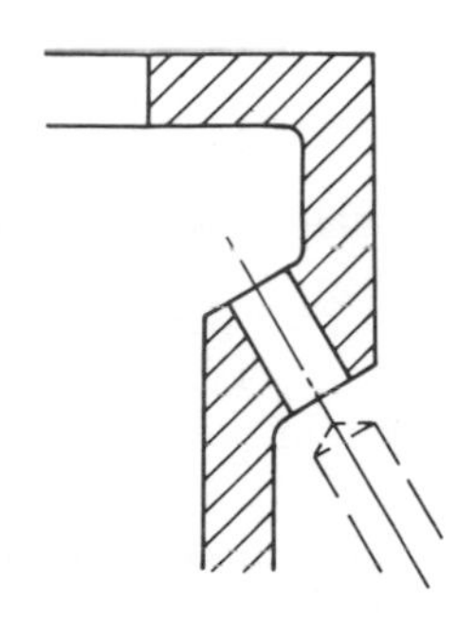

Fig. 7.17 Design for alternative tool approach

Fig. 7.18 shows an approach problem often met when surfaces are to be machined by milling. In this example the top surface of flange A is to be milled, but flanges B and C prevent the use of a roller cutter, which is the more convenient cutter to use. By raising flange A or lowering flange B a clear path for a roller cutter is obtained as shown in fig. 7.19.

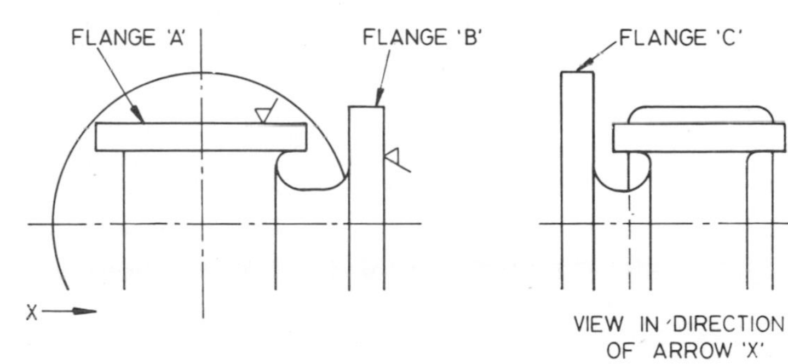

Fig. 7.18 Design with obstructed milling-cutter path

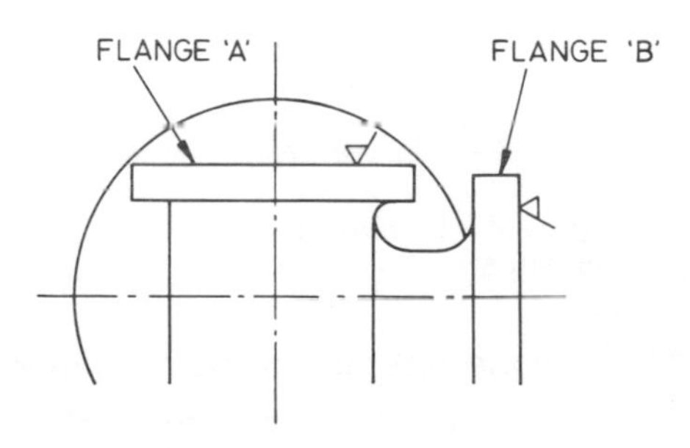

Fig. 7.19 Design with a clear path for milling cutter

When a recess is to be machined using a boring tool or boring-bar arrangement, the design must be such that the main bore is large enough to allow the tool assembly to enter and be fed out to produce the recess. Fig. 7.20 shows an unsuitable design and fig. 7.21 shows a suitable design.

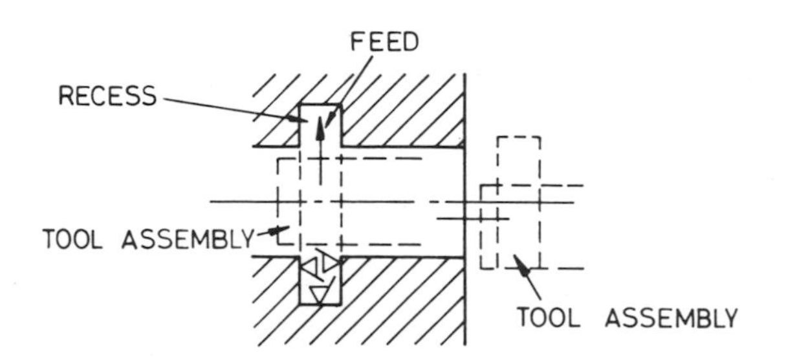

Fig. 7.20 Design with a bore that will not accept the tool assembly

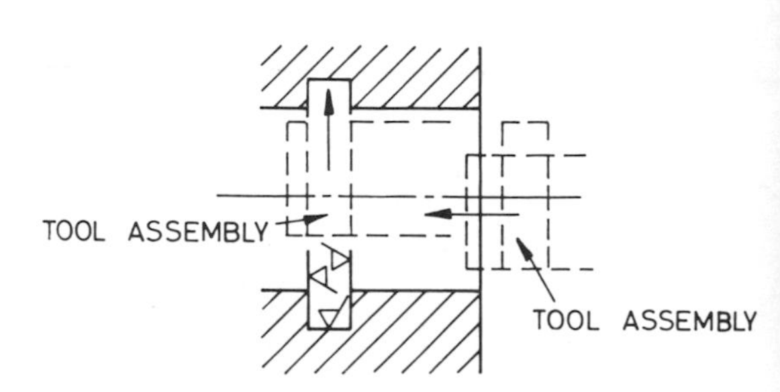

Fig. 7.21 Design with bore that will accept the tool assembly

(b) Design to allow cutting tool over-run

Except in a few instances, it is necessary to allow the cutting tool to over-run. Over-run may be necessary to sever the chip from the workpiece, to provide body clearance for the cutting tool, to obtain better cutting action or to eliminate the need to stop the cutter at a precise point.

Fig. 7.22 illustrates a blind and slotted hole. The recess at the bottom of the hole ensures that the chips will be severed from the workpiece and also allows the operator some latitude when setting the stroke of the slotting machine used to produce the slot.

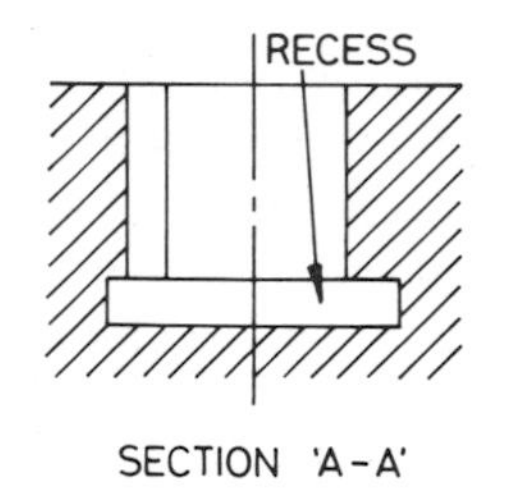

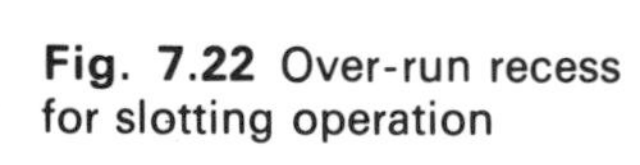

Fig. 7.22 Over-run recess for slotting operation

When a through hole is drilled, the drill must be allowed to over-run so that it breaks through and completes the hole. The design shown in fig. 7.23 does not allow an adequate over-run, but by a slight modification to the position of the angular wall, as shown in fig. 7.24, an adequate over-run is obtained.

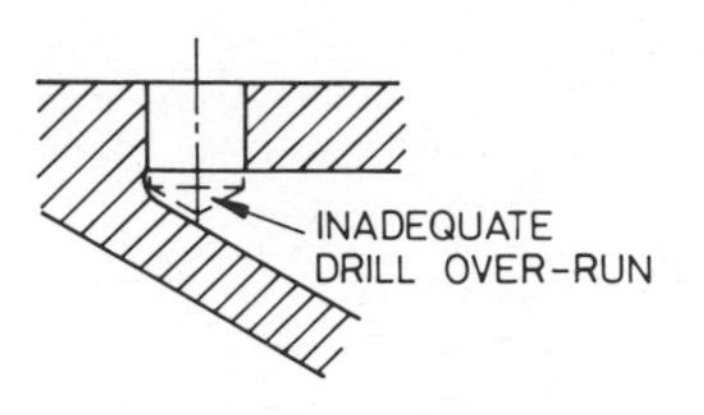

Fig. 7.23 Design with inadequate drill over-run

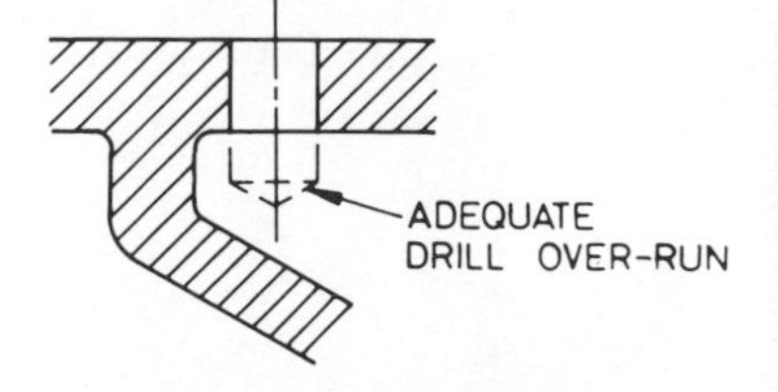

Fig. 7.24 Design with adequate drill over-run

When a shouldered, or multi-diameter, cylindrical component is ground using a traverse, a groove should be introduced so that

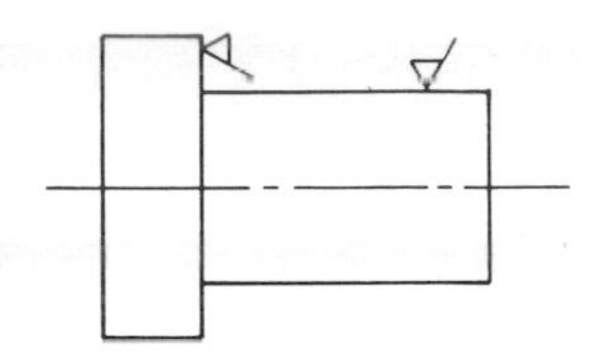

Fig. 7.25 Design without grinding over-run groove

a precise stopping point is not required and so that the best cutting action is obtained. Fig. 7.25 shows an unsuitable design and fig. 7.26 shows one that includes an over-run groove. Fig. 7.27 shows a shouldered conical component and fig. 7.28 shows how the necessary over-run is obtained by introducing a cylindrical section between the cone and the shoulder.

When a groove is milled using a side-and-face milling cutter, the relationship between the depth of the groove and the diameter of the cutter and its spacing collars must be considered. The cutter and collar diameters must be sufficiently different to make it possible to cut the groove to the specified depth. This is of special importance when the groove is to be cut close to a wall or shoulder as illustrated in fig. 7.29. In order to cut a groove to depth P and to ensure that there is adequate clearance N between the workpiece and the spacing collars, the cutter must be of a suitably large diameter. The distance L between the cutter axis and the wall is the shortest that will still give the minimum clearance M. This will affect the function of the groove because of the run-out of radius, RAD, which is controlled by the cutter diameter. A groove can be cut to full depth nearer to a wall by using an end mill as illustrated by fig. 7.30; but it must be emphasised that the teeth of such a cutter are smaller, and therefore more fragile, than those of a side-and-face milling cutter, and the feed rate is reduced accordingly. The radius RAD at the end of the groove is equal to the radius of the cutter. When the groove is to be produced by a single pass of the end mill, its width W must be the same as the diameter of the end mill, and the radius RAD must be equal to $\frac{1}{2}$ W.

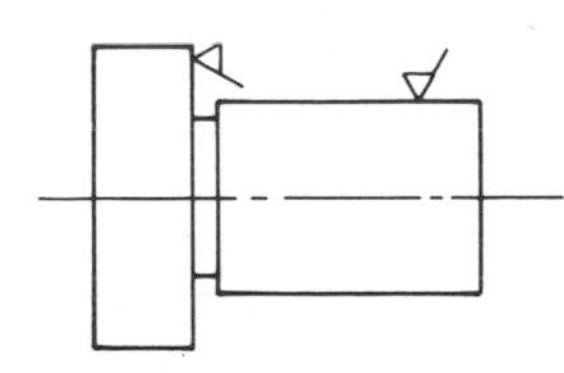

Fig. 7.26 Design with grinding over-run groove

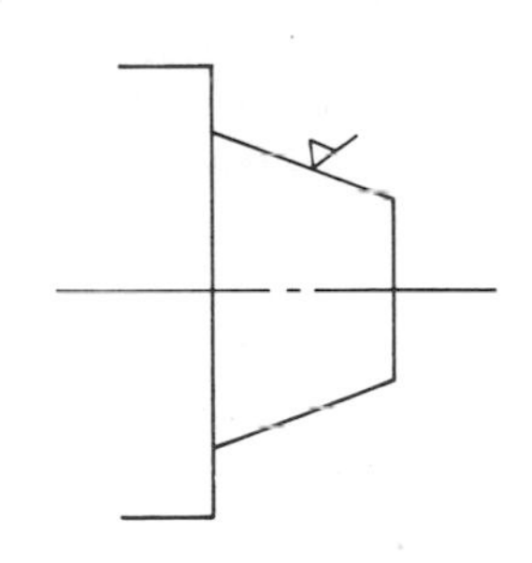

Fig. 7.27 Conical feature without over-run

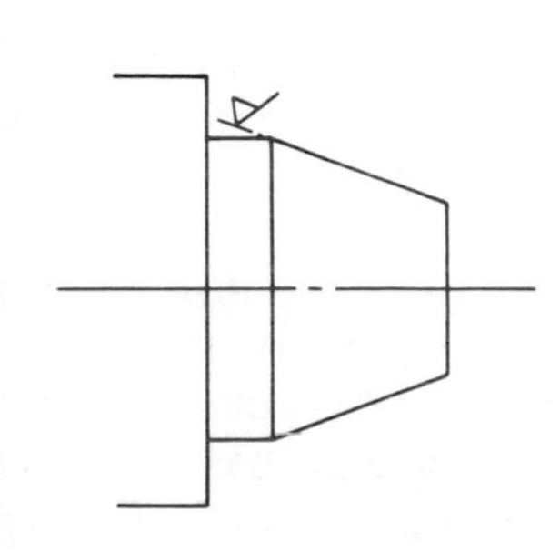

Fig. 7.28 Conical feature with provision for over-run

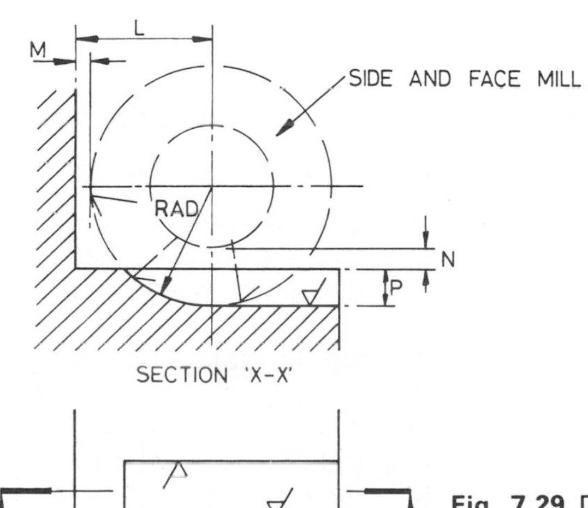

Fig. 7.29 Design of groove to be produced by a side-and-face mill

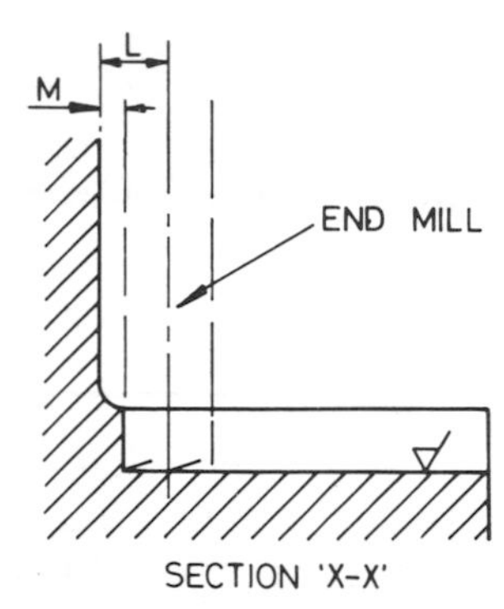

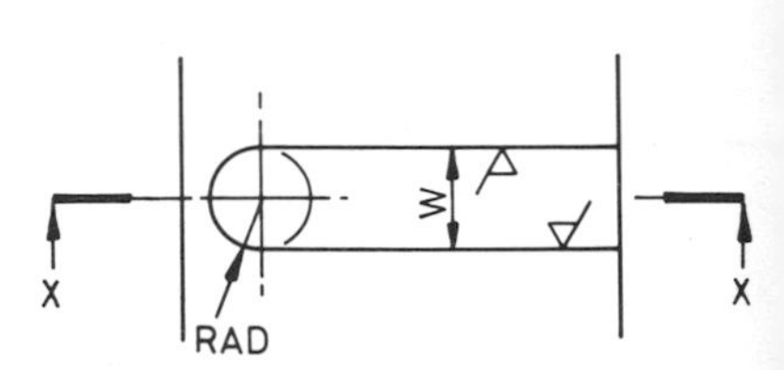

Fig. 7.30 Design of groove to be produced by an end mill

(c) Design to minimise tool deflection

A cutting tool of cantilever design will deflect during cutting if the overhang is large compared with its depth and this deflection will cause inaccuracies. The situation illustrated by figs 7.20 and 7.21 includes the problem caused by the main bore being too small to permit the use of a boring bar that would not deflect. The boring of long workpieces causes problems because of the necessarily long boring-bar overhang to reach far enough; horizontal boring machines incorporate means of supporting the boring bar at its far end so that becomes a beam, thereby reducing the deflection. The support cannot be used when a blind hole is bored and so a cored hole should, if possible, be introduced, as shown in fig. 7.31, to permit the support of the boring bar.

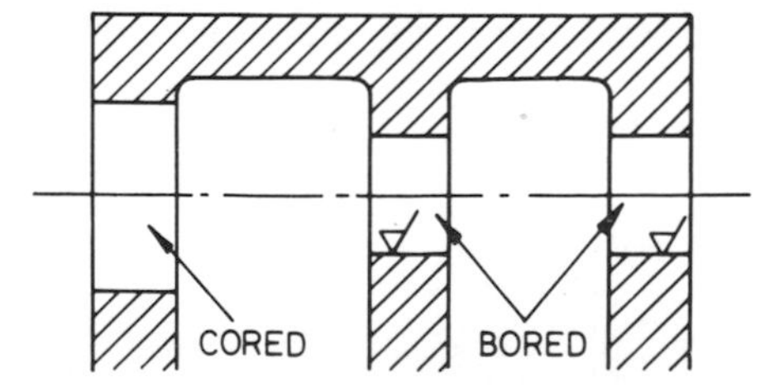

Fig. 7.31 Design to enable boring bar to be supported

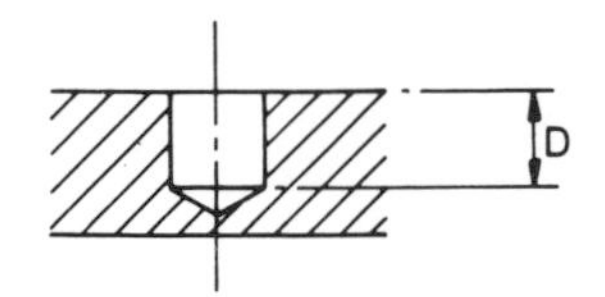

Fig. 7.32 Hole drilled to depth

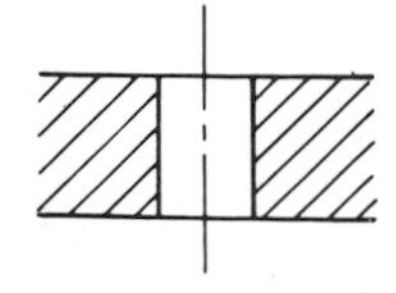

Fig. 7.33 Through hole

(d) Design to cut costs

Machining costs can be minimised by reducing the setting costs. For example the drilled hole shown in fig. 7.32 almost breaks through and the time taken to drill right through will be almost the same. Unless a blind hole is essential it would be better to alter the design to that shown in fig. 7.33, eliminating the equipment and time to set the machine to give the required depth, D.

A change of setting during an operation increases the time taken and may also lead to errors. Fig. 7.34 shows two facings whose heights differ by H. A change of setting is eliminated by making them the same height as shown in fig. 7.35.

The need to re-position the workpiece or to split an operation

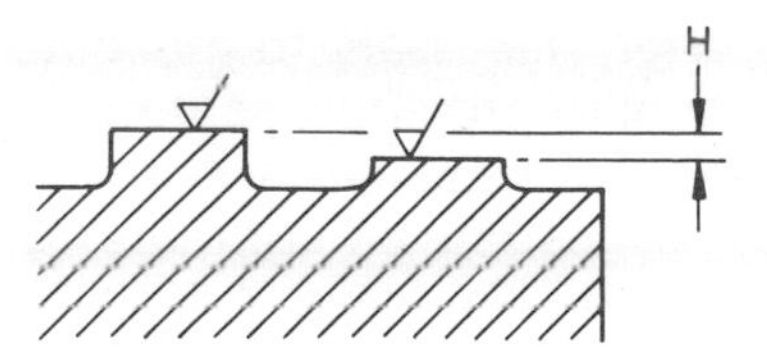

Fig. 7.34 Facings in surfaces in different planes

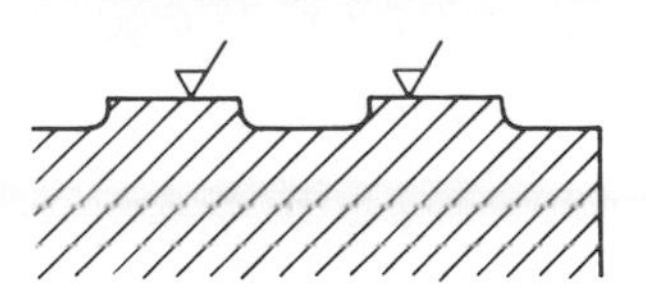

Fig. 7.35 Facings with surfaces in same plane

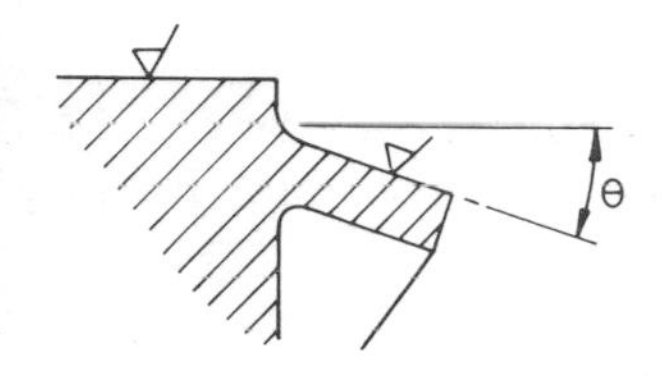

Fig. 7.36 Angular facing

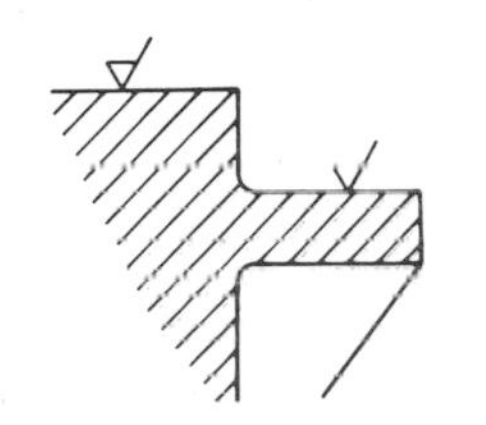

Fig. 7.37 Facing parallel to main surface

can be eliminated by making features lie in the same plane, or at least in parallel planes, as illustrated by figs 7.36 and 7.37. Similarly, holes produced by a drilling machine should be of the same diameter to avoid tool changes, have parallel axes to avoid re-positioning the workpiece, and be to the same depth to avoid the need to alter the depth control device.

Costs can also be reduced by minimising the surface area to be machined and the amount of metal to be removed from them. Fig. 7.38 to 7.43 illustrate three examples of casting design to minimise the surface areas to be machined. Minimising the area of location surfaces more nearly satisfies the principles of location because large areas are more prone to geometric errors during machining or when the part is in service.

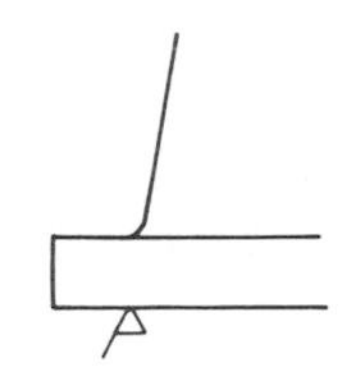

Fig. 7.38 Design with extensive seating surface

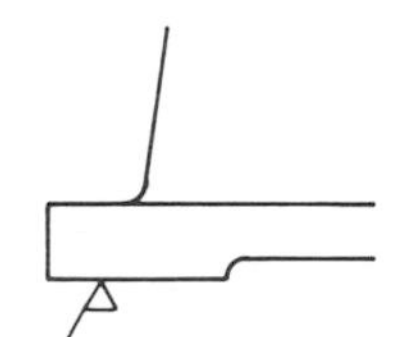

Fig. 7.39 Design with relieved seating surface

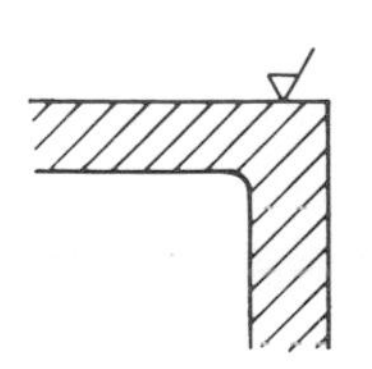

Fig. 7.40 Design with large machined surface

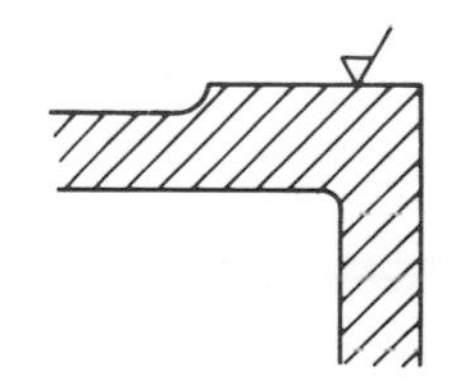

Fig. 7.41 Design with relieved machined surface

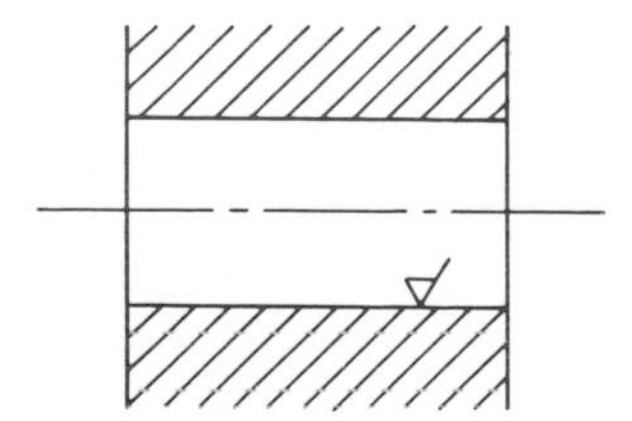

Fig. 7.42 Design with unrelieved bore

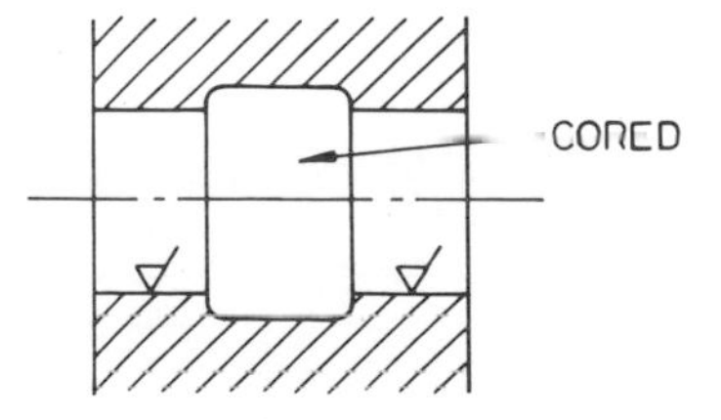

Fig. 7.43 Design with bore relieved by coring

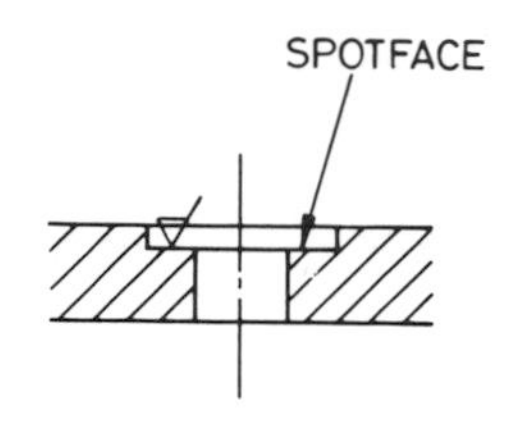

Fig. 7.44 Application of spotface

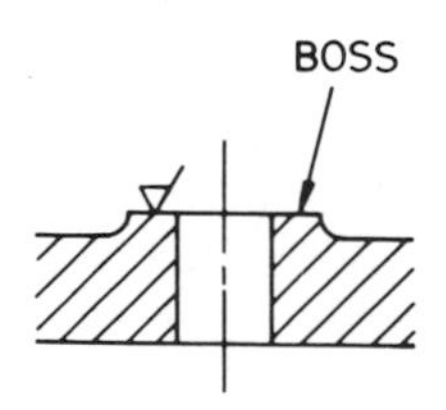

Fig. 7.45 Application of boss

It is often necessary to machine the workpiece adjacent to a drilled hole to provide a seating surface for a bolt head or other component. This can either be done by localising the machining (fig. 7.44 illustrates a spotface) or by localising the metal that surrounds the hole (fig. 7.45 illustrates the use of a boss).

The amount of metal to be removed depends upon the accuracy of the primary manufacturing process. For example casting is more accurate than forging, and die-casting is more accurate than sand-casting (to the point of eliminating the need for machining to obtain the desired degree of accuracy). The decision regarding the choice of manufacturing method must be made with due regard to other factors. For example, forging produces a stronger part than does casting, sand-casting can be used to manipulate metals of higher melting point than can die-casting, and die-casting is only economically justified when the rate of production is high and the total quantity required is very large.

8 Fastening and Joining

Components and structures which are large or complicated are produced by securing pieces of material together because other manufacturing methods, such as casting or forging followed by machining, are unsuitable. Assemblies are, in turn, produced by securing components together. Non-permanent and semi-permanent securing is usually termed *fastening* and permanent securing is usually termed *joining*. The reader should refer to a suitable text book for details of screw-thread fastening systems and joining methods. Manufactures' literature will give details of the many fasteners that are on the market.

The method used to secure the material that forms a component or structure should be selected before the initial design work is completed because it will influence the basic shape of the product. The method used to secure the components that make up an assembly should be decided at an early stage in the detail design work because it will influence both the shape of the components immediately involved and that of components in the vicinity of the joint.

8.1 The factors that influence the choice of securing method

- The degree of permanency that is required.
- The size and shape of the pieces or components to be secured.
- The required strength of the structure, component or assembly.
- The composition, mechanical properties and physical properties of the materials involved.
- The limiting conditions—for example, the effect of heat upon

the material, or upon components in the vicinity.
- The appearance of the product around the joint.
- The allowable cost of the fixing operation.
- The ease with which the components or parts must be separated for maintenance or repair.

8.2 Fastening systems

The principal fastening systems can be classified as:
(a) systems that use screw threads;
(b) fasteners;
(c) pins;
(d) rivets.

(a) Fastening systems that use screw threads

Pieces of material can be secured, and assemblies produced using nuts and bolts when both the parts can be drilled to a suitable clearance size and when there is sufficient space for the bolt to be fed into the hole, for the washer and nut to be placed on the bolt, and for spanners to be used to tighten or release the parts.

When one part is thick enough and made of a material that is strong enough to be threaded, a *set screw and threaded part* system can be used. This system is used where there is no room for a nut, or when two spanners would be awkward to use; but it requires care at the assembly stage because a crossed thread would damage a major component, possibly beyond repair.

A stud and nut system can be used if one part can be threaded to take the stud, or if the stud can be introduced when that part is cast (when metal parts are produced by die-casting) or moulded (when plastics parts are produced by injection moulding). This system is useful when the stud is required to locate the second part and when a single-spanner tightening system is required. Introducing the stud at the casting or moulding stage is done when the material is not strong enough to be threaded, and has the advantage that the threading operation is unnecessary. Care must be taken when the fastening is done because damage to the stud would scrap the component to which it is attached. When studs or threaded holes are required in sheet metal, fasteners secured in a similar way to rivets can be used; alternatively, they can be attached by projecting-welding.

Screw thread systems are particularly useful when the parts will need to be separated for maintenance or repair without damage to the main securing components. Unfortunately they are often unsuitable for external application because of their somewhat untidy appearance. These systems are prone to loosening by vibration, but this can be overcome by using self-locking nuts, lock nuts, or locking devices such as tab-washers, spring washers or locking wire.

Self-tapping screws are case-hardened screws that can produce their own mating thread in the parent material whilst they are securing the parts together, and so eliminate more costly systems, such as the nut and bolt system. There are a number of self-tapping screws available; the more commonly used ones produce threads in holes that have been previously produced, but other types produce their own holes and their mating thread.

(b) Fasteners

A large number of fasteners are commercially available to rapidly attach panels to sheet metal, join sheet metal panels together and to attach components, such as pipe clips, to other components. Figs 8.1, 8.2, and 8.3 show some typical fasteners; these require a hole to be produced in the parent component before fastening.

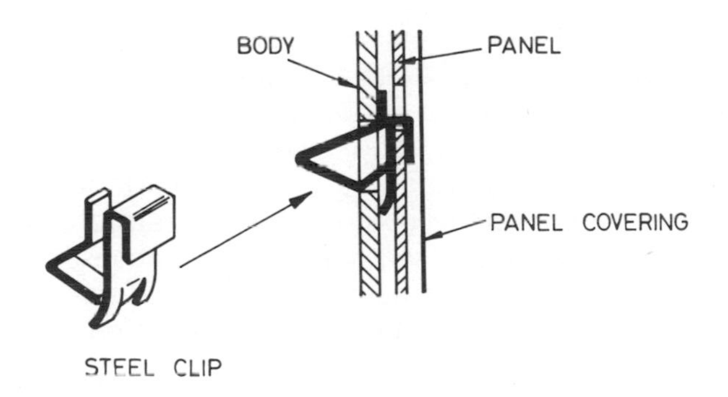

Fig. 8.1 Spring clip for fastening a panel to a body

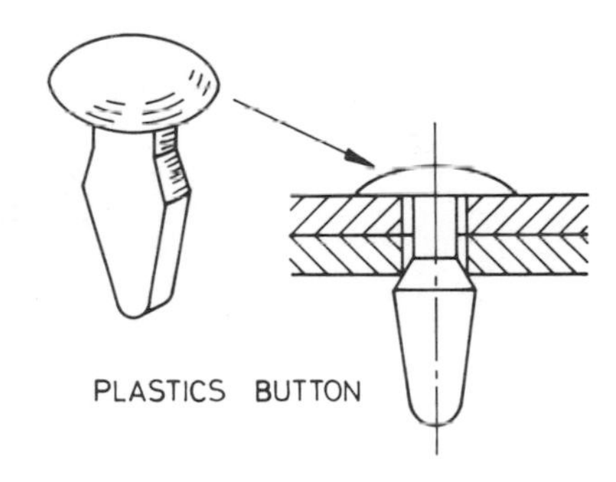

Fig. 8.2 Plastics button to join panels

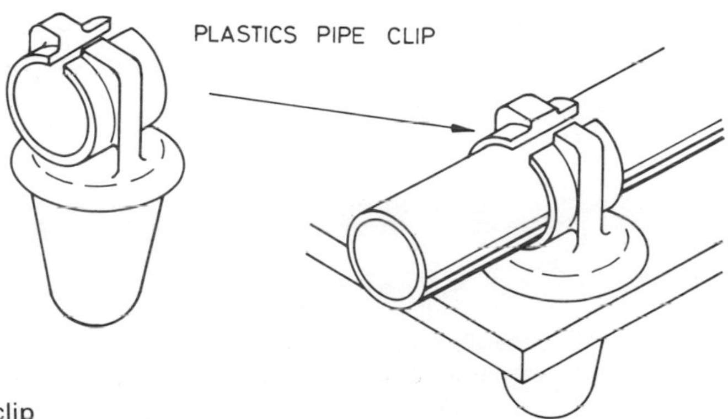

Fig. 8.3 Plastics pipe-support clip

(c) Pins

A pin is used to locate and secure a component in direction, or directions, normal to its axis. When plain cylindrical pins are used the hole and pin must be accurate to ensure positive location and retention. The degree of accuracy that is necessary can be reduced by using taper pins, but then taper reaming is required. Roll pins (see fig. 8.4) do not require high accuracy because they compensate for clearance by springing open.

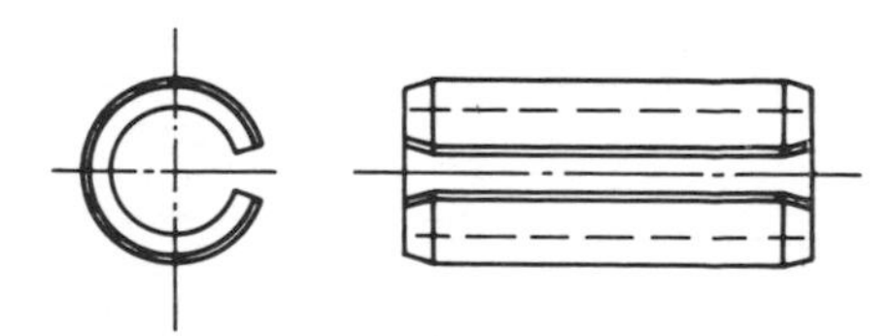

Fig. 8.4 Roll pin

(d) Rivets

Rivets are used when a semi-permanent securing system is required. They are cheaper and quicker to apply than are threaded devices, are hard to tamper with and are not loosened by vibration. When properly installed they provide a strong joint and yet enable replacement of parts because they can be removed by drilling.

When there is access to both sides a solid rivet can be used. A variation of the solid rivet, called a tubular rivet, which is a solid rivet drilled almost up to its head, requires less force to clinch. It is particularly suitable for securing thin sheets or components that would buckle if conventionally riveted. A wide range of blind rivets are available, and may be used where there is access to only one side. A blind rivet is clinched using a special gun that pulls a mandrel through it—the mandrel being designed to snap when the required clinching conditions are reached.

8.3 Joining methods

Joining is generally used when a permanent securing system is required. The principal joining methods are:

(a) soldering;
(b) brazing;
(c) welding;
(d) bonding using structural adhesives.

(a) Soldering

Soldering implies joining using an alloy that melts at a lower temperature than do the metals to be joined, forms a liquid solution at their surfaces, and solidifies to form a continuous structure across the joint. A typical solder melts at about 180° C and does not adversely affect the material or components being joined. Parts can be separated by reheating to melt the solder. The joint is fairly weak, but soldering can be used to secure other joining systems such as wire-connectors. The solder will only form a liquid solution at the surface of the metal being joined if the surfaces are clean and free from oxide, and so the process is done using a suitable flux.

Soldering may be done using a soldering iron; or, when the production rate is high, by assembling the parts to be joined with a solder-flux mixture at the joint, and simultaneously heating several assemblies. When the latter system is to be used, the parts should be designed accordingly.

(b) Brazing

This system is basically the same as soldering but it uses a solder which melts at a higher temperature. There are a large number of brazing solders and silver solders. Some of these start to melt at temperatures as low as about 625° C while those with the highest melting temperatures start to melt at about 870° C. Brazing produces a stronger and tougher joint than does soldering, but the effect of the higher temperatures upon the strength of the metals to be joined must be considered.

(c) Welding

Welding systems can be broadly classified as (i) fusion-welding and (ii) pressure-welding. Fusion-welding involves heating the metal at the joint to melt it and usually a filler metal, which is often similar to that being joined, is introduced. Pressure-welding produces a joint by applying pressure—usually when the metal is solid.

Fusion-welding is similar to casting. Problems such as oxidation, variation of structure and workpiece distortion must be taken into account both when selecting the specific system to be used and when designing the product. Fabrication by welding, as an alternative to production by casting or by working, and finishing by machining, has become important during recent years. This is of special importance in the aerospace industry where, to reduce the mass of engines, components are of thin section. The manufacture of these components from castings or from forgings involves the removal of a considerable amount of metal. This increases the costs because of the time it takes, is a waste of metal, and induces stresses or causes distortion unless special costly metal-removal processes are used.

Variations of electric-arc welding have been introduced to overcome the oxidation problem. New processes, such as

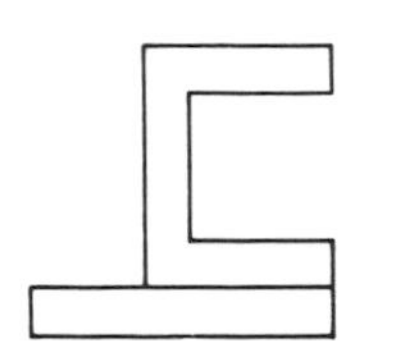

Fig. 8.5 Unsuitable arrangement for spot-welding

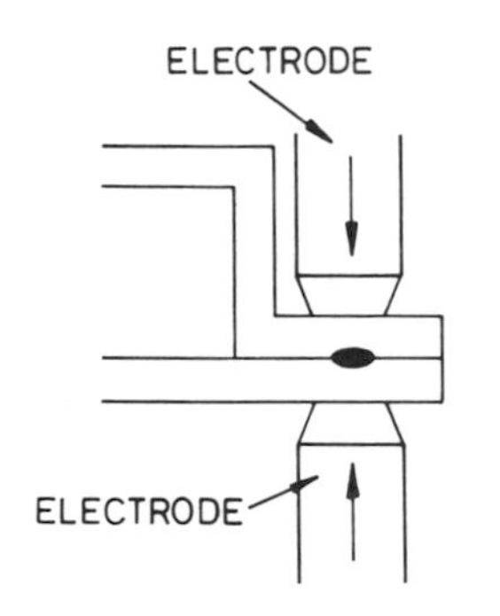

Fig. 8.6 Suitable arrangement for spot-welding

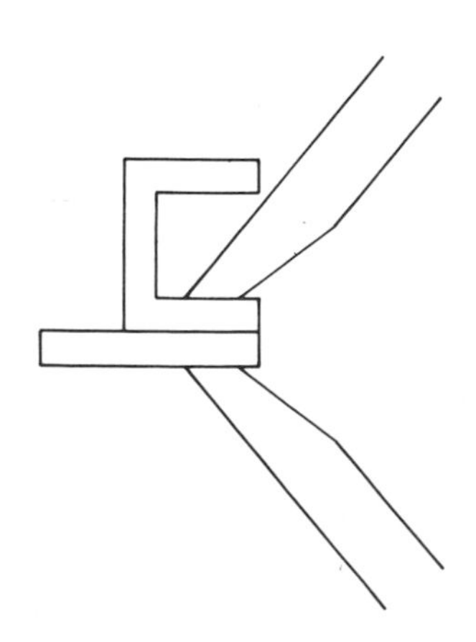

Fig. 8.7 Offset electrodes

electron-beam welding, have been developed to enable better quality welds to be obtained and to enable awkward metals to be joined. The reader is advised to study the various fusion-welding processes and be aware of their characteristics if he or she is likely to be involved with design for welding.

Resistance-welding processes involve passing an electric current at low voltage through two pieces of metal that are in close contact so that the heat that is thereby generated at their interface causes the metal to locally melt. The weld is controlled during its solidification and during its cooling in the solid state by maintaining pressure across the interface—resistance-welding is therefore a combination of fusion-welding and pressure-welding. Spot-welding (which is a resistance-welding process) is particularly suitable for joining sheet metal and lends itself to welding by robots. In this system, a high current density is obtained by applying force over a small area during current flow by using shaped electrodes. The diameter of the spot is approximately the same as that of the electrodes (which should be approximately $5\sqrt{t}$—where t is the thickness of the sheet).

When designing for spot-welding, the access of the electrodes is of prime importance. In order to concentrate the resistance, and hence concentrate the heating, at the interface, the electrodes must be of high electrical conductivity and the contact pressure must be high. The electrodes are made of copper—which is comparatively weak. It is therefore best if they are presented in line from both sides of the workpiece to prevent their deflection. See figs 8.5 and 8.6.

Offset electrodes, as shown in fig. 8.7, can be used for spot-welding material up to 3 mm thick, but thicker material requires high forces and the electrodes will tend to skid across the work. This produces joints of poor quality and causes rapid wear of electrodes.

The weld should not be too close to the edge of the metal otherwise the edge will be distorted and the weld will be of poor strength as shown in fig. 8.8. The distance from the edge of the metal to the weld should be at least 1.5 d (where d is the diameter of the weld). When flanged parts are to be spot-welded, care should be taken to ensure that neither electrode touches the upturned parts because shunting will occur if they touch and a poor spot-weld will be obtained (see fig. 8.9).

Shunting though a previous weld will occur if the weld pitch is too small (see fig. 8.10). The weld pitch may be increased but this may result in an unsatisfactory joint. The alternative is to adjust the conditions to compensate for the reduction of heat caused by the shunting, although the first weld of each run will then be 'too hot'.

Poor quality welds are often caused by a poor fit-up because some of the welding force is taken up in collapsing the faulty

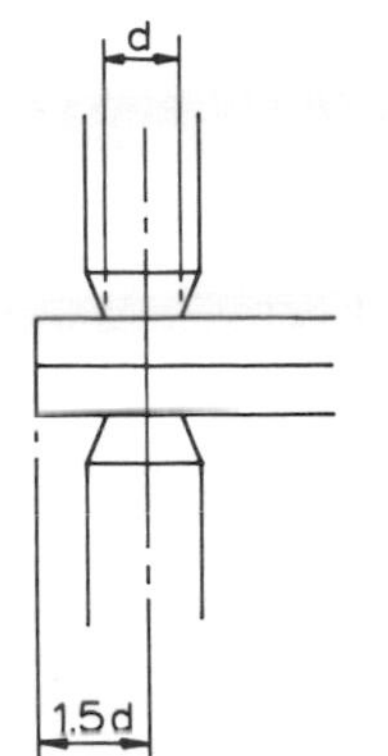

Fig. 8.8 Spot-welding at edge

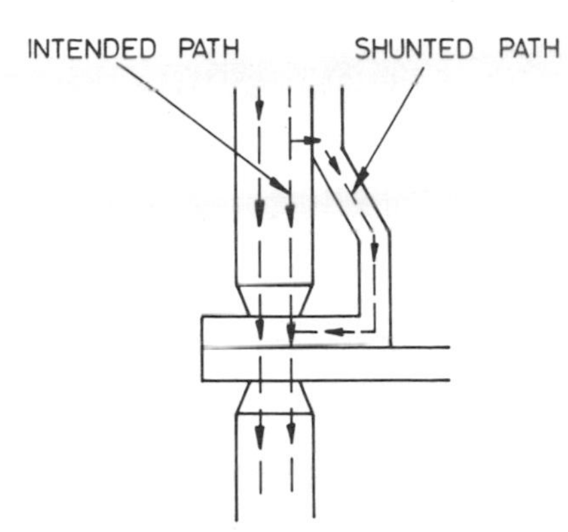

Fig. 8.9 Shunting caused by touching a part

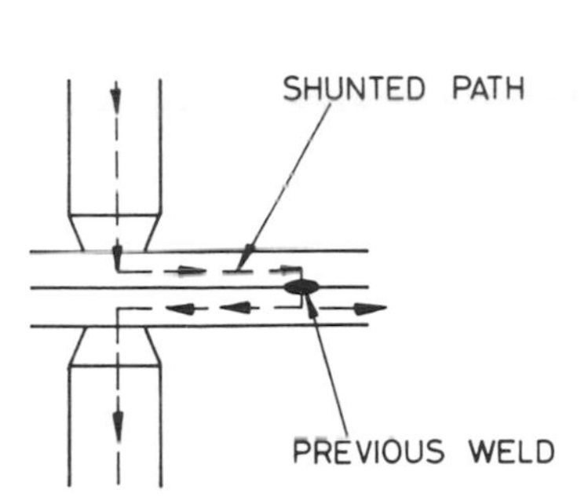

Fig. 8.10 Shunting caused by incorrect weld pitch

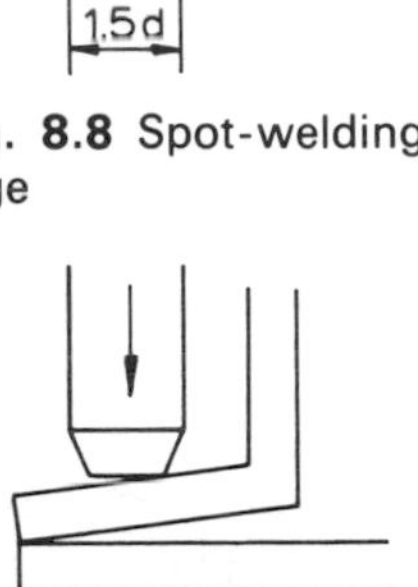

Fig. 8.11 Bad fit-up

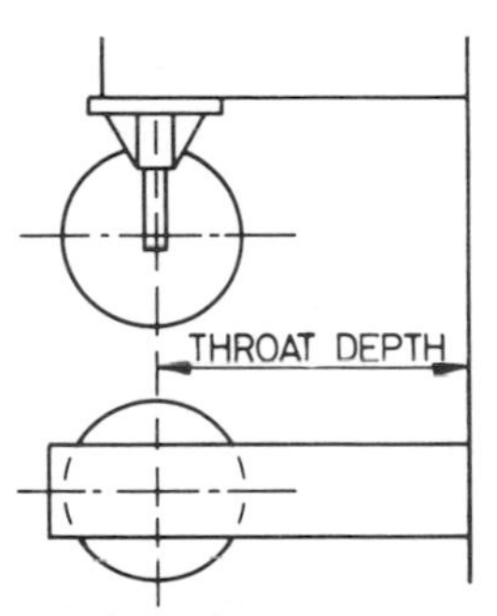

Fig. 8.12 Welding-head configuration

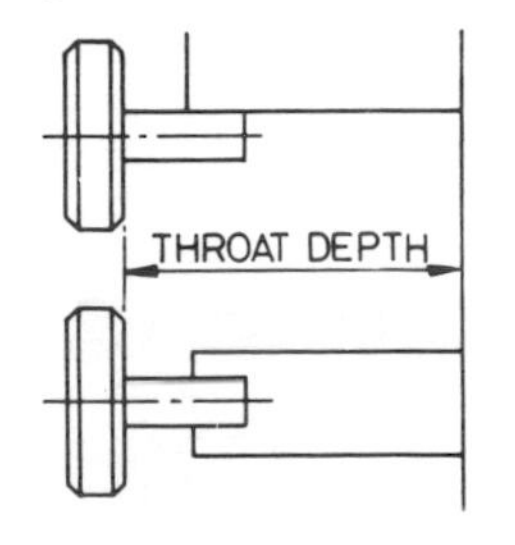

Fig. 8.13 Welding-head configuration

components (see fig. 8.11). Poor fit-up is usually caused by spring-back during presswork rather than by poor design; the usual solution is to employ a jig system to enable the operator to easily identify a poor fit-up.

When the suitability of metal for spot-welding is being considered, its electrical conductivity (i.e. its resistance) must be taken into account because the heat that is generated depends upon the resistance. The thermal conductivity must also be considered because loss of heat from the heated zone due to a high thermal conductivity may result in a poor quality spot-weld. The combination of high electrical conductivity and high thermal conductivity may be overcome by using a high current and a short welding time (less than 0.2 s). Metals of low thermal conductivity are awkward because the heat at the weld is not dissipated and the metal may boil and molten metal may be expelled. This may be overcome by using a high current, high electrode forces and a long welding time (0.2 to 0.6 s). The surfaces to be welded must be free from rust and heavy scale that would act as an insulator and the removal of the oxide that forms on the surface of some metals (such as aluminium) must be considered because its intermittent breakdown produces local hot spots followed by a period of no heating.

Seam-welding produces a number of overlapping spot-welds using profiled wheel electrodes that produce a continuous force, and applying an intermittent current. The design considerations are similar to those when spot-welding but the welding current values are different because of the shunting effect of the overlapping spots. Figs 8.12 and 8.13 show the two basic welding head configurations. The product should be designed so that it can be accommodated by the seam-welding machine that is available.

It often seems prudent to design parts so that they are self-locating. Unfortunately this will cause shunting as shown in fig. 8.14. The design shown in fig. 8.15, which is not self-locating, is better because it does not cause shunting.

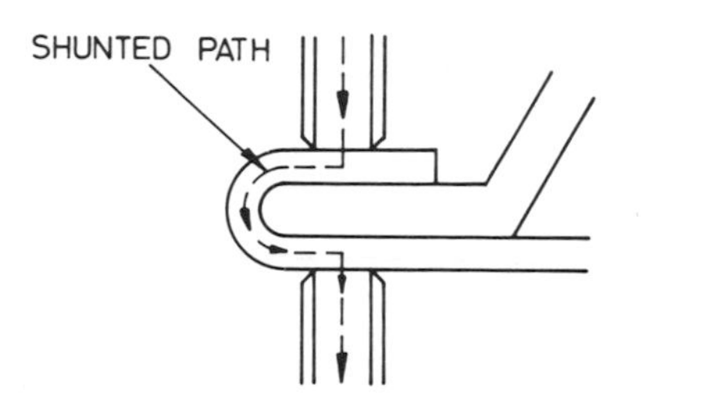

Fig. 8.14 Shunting caused by location

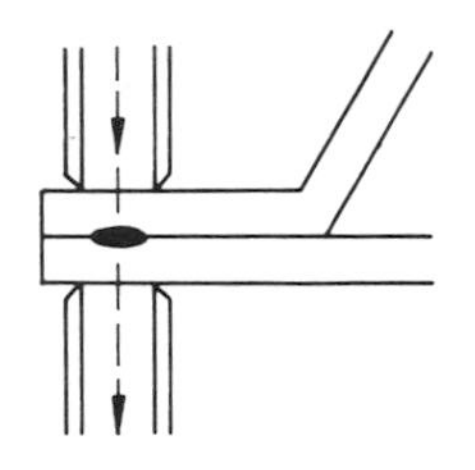

Fig. 8.15 Prevention of shunting

Projection-welding is a resistance-welding system in which the current flow is localised by the geometry of the parts rather than by the shape of the electrodes (see fig. 8.16). The larger electrodes will have a longer life than those used when spot-welding. The dimple, which is produced by presswork, must be rigid enough to prevent its collapse under the electrode force before the passage of the current and be of large enough diameter to produce a weld of adequate size. The line of action of the electrode force should, if possible, be perpendicular to the surface of the parts being welded. But if the shape of the parts makes this impossible the dimples should be elongated in the direction of the movement of the parts so that their movement does not cause the weld nuggets to be completely sheared (see figs 8.17 and 8.18). Projection-welding can also be used to weld studs to sheets as illustrated by fig. 8.19.

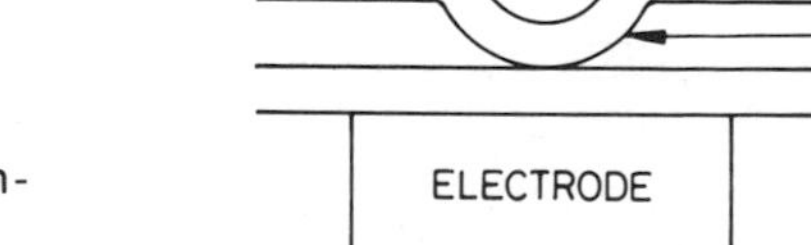

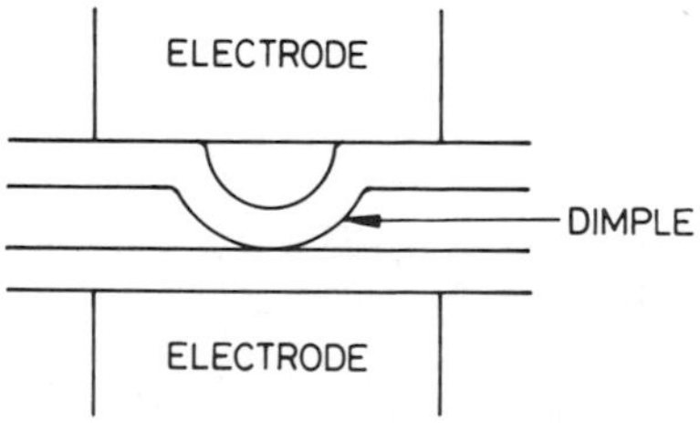

Fig. 8.16 Projection-welding

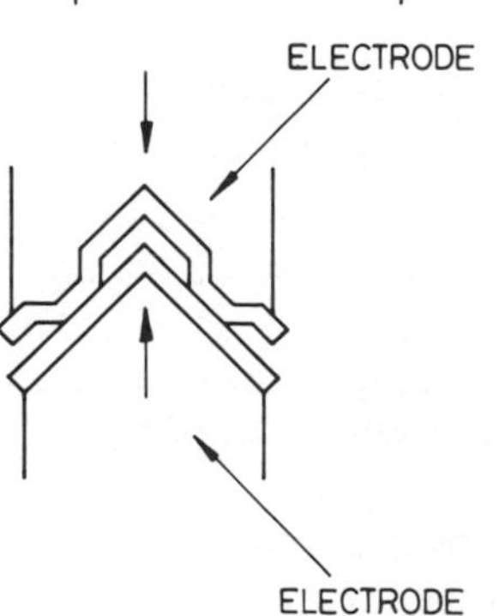

Fig. 8.17 Design to prevent weld shearing

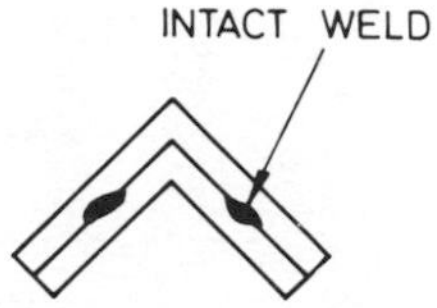

INTACT WELD

Fig. 8.18 Result of suitable design

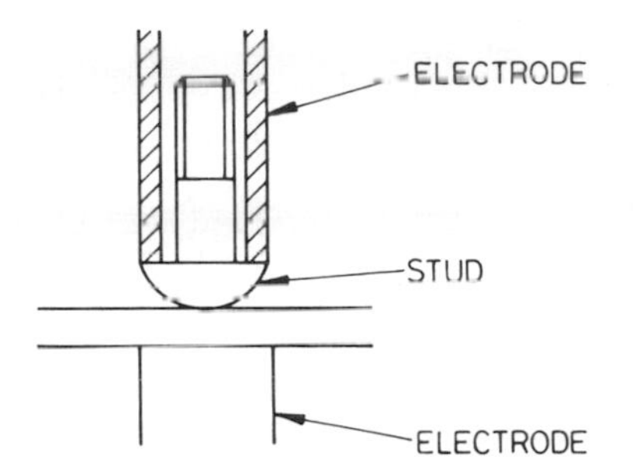

Fig. 8.19 Welding of a stud by projection-welding

Friction-welding is a true pressure-welding system in which the metal remains in the solid state. The heat needed to produce the weld is generated by rotating one part against another part which is held firmly under axial pressure. The process is very rapid and, because it is mechanical, unskilled labour can be used and the process fully automated. Friction-welding enables nearly similar, and also dissimilar, metals to be joined. One part must, in the region of the weld, be round or have an axis of symmetry in rotation, and the other part must be capable of being adequately restrained against the applied torque. The size of the parts is limited mainly by the size of the machine.

Diffusion bonding is a pressure-welding process in which bonding occurs over a period of time during which atomic diffusion takes place. A bond which involves only minimal plastic deformation is produced. The bond is produced by a combination of time, pressure and temperature; and if, for example, the pressure cannot be high because the parts to be joined are too weak, a higher temperature for a longer time can be used to compensate. There must be a good fit-up between the surfaces to be bonded and they must be clean and free from contamination. The parts are normally transferred to the bonding atmosphere, usually an inert gas or vacuum, within a few minutes of cleaning and the pressure applied. The bond is almost undetectable and a wide range of awkward materials can be joined; but the need to use controlled atmosphere pressure vessels, to produce an accurate fit-up, and to have a satisfactory surface preparation limits the process on a commercial scale.

(d) Bonding using structural adhesives

Unlike traditional joining methods, which produce local stress points, correctly located and well designed bonded joints have a loading over the whole adhesive area, enabling lighter gauge metals to be used. The product will have a smooth contour in the region of the joint, fixing holes are not required, the joint will have a good fatigue resistance and the joint is sealed as well as bonded. Also metals can be bonded to dissimilar metals and to non-metals, and because most structural adhesives are electrical

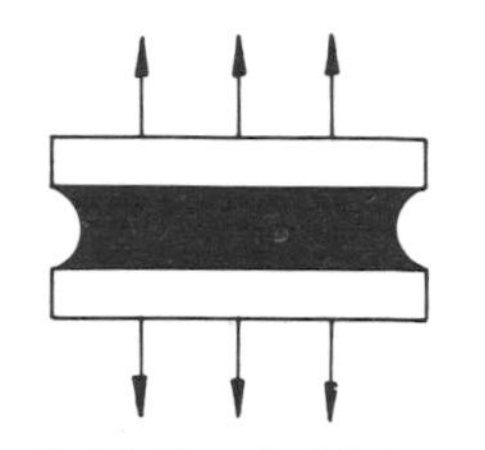

Fig. 8.20 Bonded joint in tension

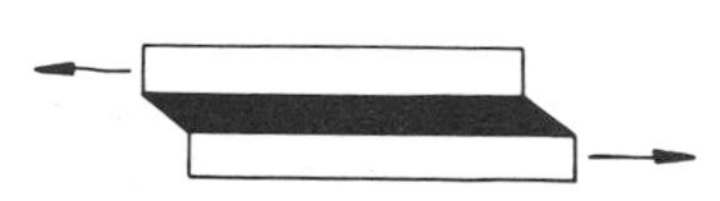

Fig. 8.21 Bonded joint in shear

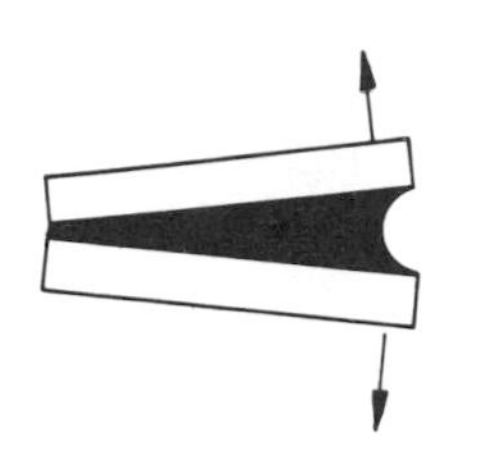

Fig. 8.22 Bonded joint in cleavage

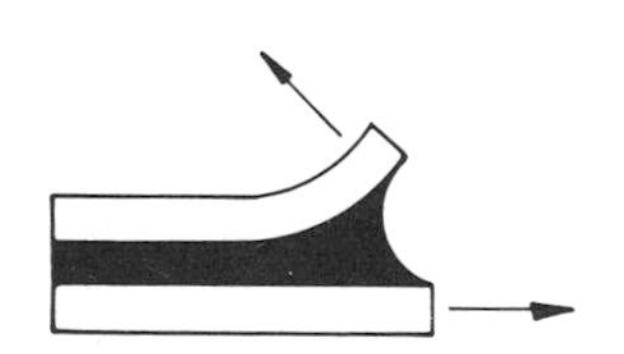

Fig. 8.23 Bonded joint in peel

insulators there will be no electrical continuity across the joint; the electrical insulation prevents corrosion of dissimilar metals caused by galvanic action. The curing of adhesives, which develops their strength, requires minimal or no heating and so distortion and structural changes caused by heat do not occur. The strength of a bonded joint depends upon the strength of the adhesive and the design of the product, but it is essential that the parts to be joined are clean and the assemblies are jigged.

Structural adhesives can be classified as (i) thermoplastics, (ii) thermosets and two-part polymers and (iii) elastomers. Thermoplastics are supplied in the form of tapes, rods, pellets etc. which are applied when warmed above their softening point and cured at room temperature; they soften when re-heated and may be prone to creep under stress. Thermosets and two-part polymers solidify as a result of chemical reaction or the application of heat; when set they cannot be re-softened and re-solidified by heat cycling. In general they provide higher bond strengths than do other adhesives and have a good resistance to creep. Elastomers are based on natural or synthetic rubbers and have a low strength but high flexibility; they are used to bond rubbers.

Figs 8.20 to 8.23 show the effect of subjecting a bonded joint to tension, shear, cleavage and peel. When the joint is in tension or shear, the stress is applied over the whole of the joint area: when stressed in cleavage, one end of the joint is under great stress, and the other is not in stress. When in peel stress, there is even less adhesive contribution to the joint strength than when in cleavage stress because the stress is confined to a narrow area at the edge of the joint. Assemblies to be bonded should be designed so that the principal service stresses are either tensile or shear, and joints involving cleavage and peel stresses should be avoided. The width of the joint is important because the strength of the joint increases linearly with joint width but increased joint length has little effect.

9 Corrosion

With the exception of the noble metals, such as gold and silver, metals occur naturally in an oxidised form and must be extracted from their ores. When they are in their metallic form, metals are unstable and tend to revert to their oxidised form. If this process is allowed to proceed unchecked, metals will become corroded.

9.1 Factors that govern the rate of corrosion

The rate of corrosion depends upon several factors which, as shown in fig. 9.1, can be classified as:

(a) the characteristics of the metal or metals;
(b) the characteristics of the environment.

(a) The characteristics of the metal or metals

It can be assumed that every metal has a tendency to dissolve or corrode. When it does so it discharges positively-charged particles, called *ions*, leaving it in a more negatively-charged condition. The charge on the metal caused by this spontaneous corrosion is called a *negative potential*. In the electrochemical series (see table 9.1) metals are arranged according to their standard electrode potentials (numerically expressed with respect to that of hydrogen, which is taken to be at zero). Although a metal's tendency to corrode may be relatively slight, it may be aggravated by the conditions in which it is used.

If a metal is in contact with a dissimilar metal in an electrolyte there will be a current flow between the two metals. This results in the more rapid corrosion of the metal at the higher (more

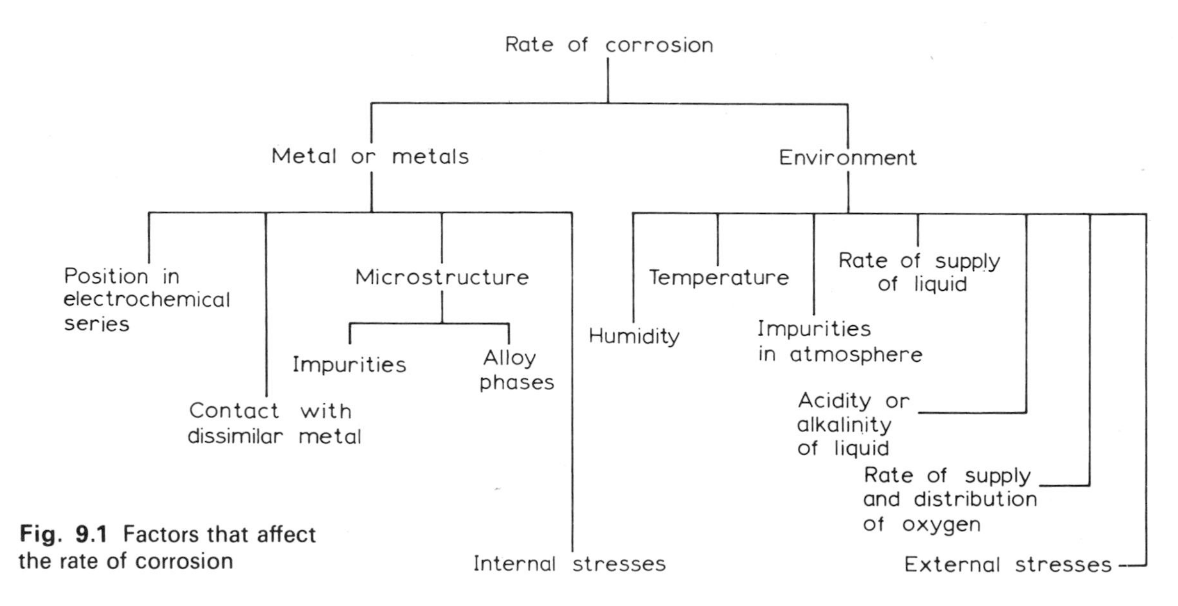

Fig. 9.1 Factors that affect the rate of corrosion

negative) end of the electrochemical series and a retardation of the corrosion of the metal at the lower end of that series. The metal that is at the higher end of the series is called the *anode* and that at the lower end is called the *cathode*. The rate of corrosion of the anode generally increases with the relative area of the adjacent cathode. This galvanic corrosion also occurs when the metal itself contains more than one phase (a *phase* is part of a chemical system with distinctive characteristics) such as an impurity in a commercially pure metal. An additional phase will be produced when metals that do not form a solid solution are

Table 9.1 Electrochemical series of metals

Metal	Standard electrode potential (volts)
Sodium	−2.71
Magnesium	−2.37
Aluminium	−1.66
Zinc	−0.76
Chromium	−0.56
Iron	−0.44
Cadmium	−0.40
Nickel	−0.25
Tin	−0.14
Lead	−0.13
Hydrogen	0.00
Copper	+0.34
Silver	+0.80
Platinum	+1.20
Gold	+1.50

alloyed (a solid solution is formed when the constituent atoms share the lattice arrays that form the grains, and the structure resembles that of a pure metal).

Internal stresses set up during casting or cold-working often cause corrosion because they produce local differences by distorting the lattice arrangement.

The cathode–anode relationship between various metals and alloys in a specific electrolyte is indicated by the galvanic series for that electrolyte (or medium) as illustrated by table 9.2. It will be seen that 18/8 stainless steel appears twice in the galvanic series in sea-water, where it is described once as being *passive* and once as being *active*; this is because it is capable of forming a protective film on its surface, when it is said to be passive. This film enables the steel to be immersed for short periods in sea-water, but when immersed for long periods the film breaks down, the steel becomes active, and it starts to corrode.

Table 9.2 Galvanic series in sea-water

	cathode
Titanium	
Monel	
Passive 18/8 stainless steel	
Silver	
Nickel	
Cupro-nickel	
Aluminium bronze	
Copper	
Brass	
Active 18/8 stainless steel	
Cast iron	
Aluminium	
Zinc	
Magnesium	
	anode

(b) The environment

As already stated, the rate of oxidation of a pure metal is comparatively low. This rate is increased as the temperature is raised. Corrosion at elevated temperatures (known as *dry corrosion*) is a special probelm in exhaust systems (and to a greater degree in gas-turbine engines) because the oxidation at high temperatures is accompanied by attack from the exhaust gases. The humidity of the atmosphere must be considered because the moisture acts as an electrolyte. Extensive observations on the relation between the rate of corrosion and the atmospheric conditions show that the rate is low in dry rural locations, high in marine locations and in industrial locations, very high in locations that are both industrial and marine, and extremely high in tropical marine locations.

The rate of supply and distribution of oxygen also influences the corrosion rate, even when the metal is pure. The aerated part is protected and becomes the cathode and the non-aerated part becomes the anode.

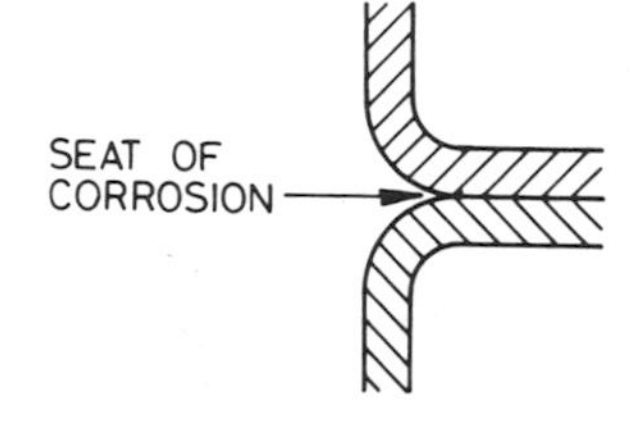

Fig. 9.2 Poor design allowing moisture to enter

9.2 Methods of preventing or minimising corrosion

Corrosion can be prevented or minimised in several ways. Some of these are simple and inexpensive; others are costly. The method adopted depends upon its cost compared with the degree of corrosion resistance that is necessary.

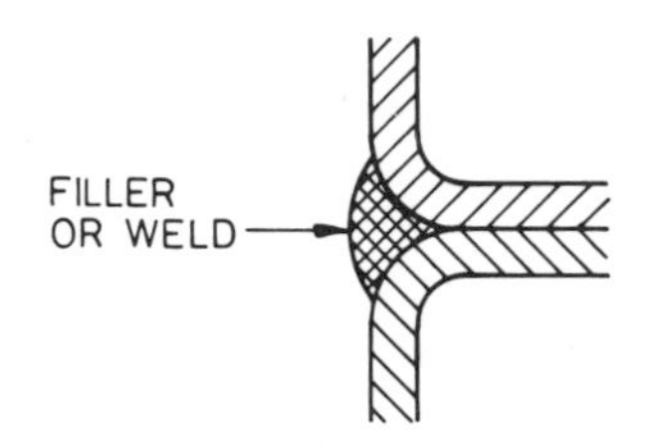

Fig. 9.3 Improved design

(a) Design to reduce corrosion

Accumulation of water can be prevented by designing so that it cannot enter (see figs 9.2 and 9.3) or designing so that it can escape (see figs 9.4 to 9.8). Galvanic corrosion can be prevented by introducing an electrical insulator between dissimilar metals as shown in figs 9.9 and 9.10.

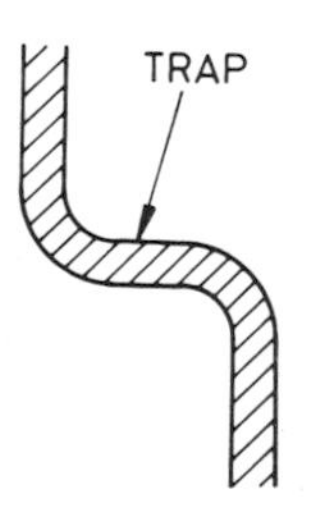

Fig. 9.4 Ledge causing moisture to be trapped

(b) Use of a corrosion-resisting material

Although, in general, a pure metal is more corrosion-resistant than its alloys, it may not be suitable either because of its high cost or because it is too weak. As already explained, a combination of acceptable strength combined with a good corrosion-resistance is obtained when metals form a solid solution when alloyed. Alloys of copper and nickel, alloys of nickel and chromium, and austenitic stainless steels (those with about 8% nickel and 18% chromium) are typical solid solution alloys. Some metals develop a surface film that gives them protection against corrosion; for example aluminium does not corrode as rapidly as its position in the electrochemical series suggests because the product of initial corrosion is a film that protects the surface from continued corrosion. When certain metals are introduced into an alloy they produce corrosion-resistance by developing such a film; for example, if more than about 10% chromium is introduced into a steel, corrosion-resistance is produced. As indicated in table 9.2, the passivity produced by the film may only be for a short period.

The problem of corrosion can often be overcome by using a non-metallic material, such as a plastics material, but the use of such materials is often limited by their low strength and by manufacturing difficulties.

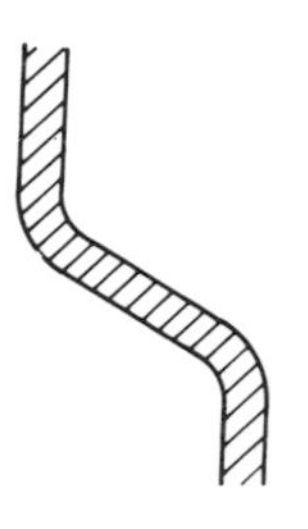

Fig. 9.5 Improved design

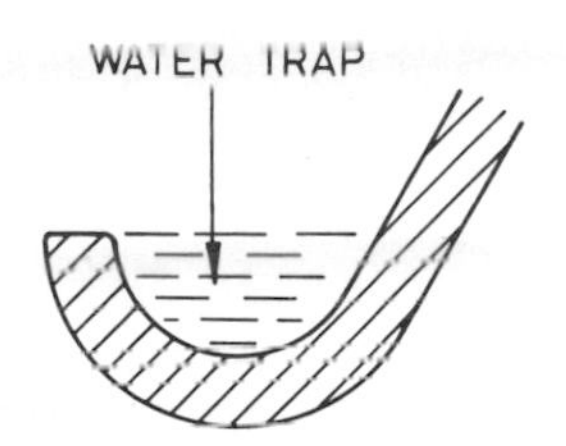

Fig. 9.6 Poor design producing a water trap

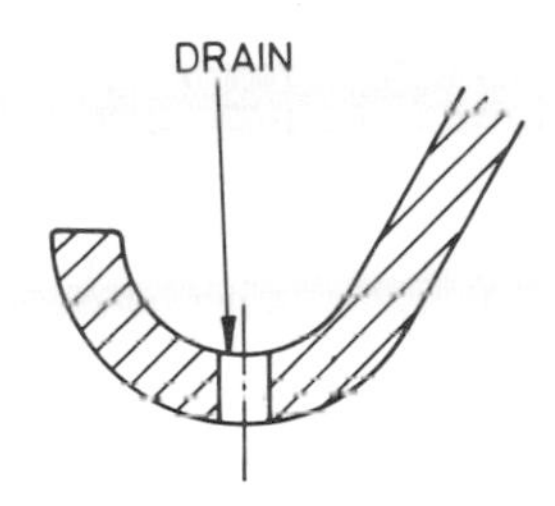

Fig. 9.7 Improved design

Fig. 9.8 Good design

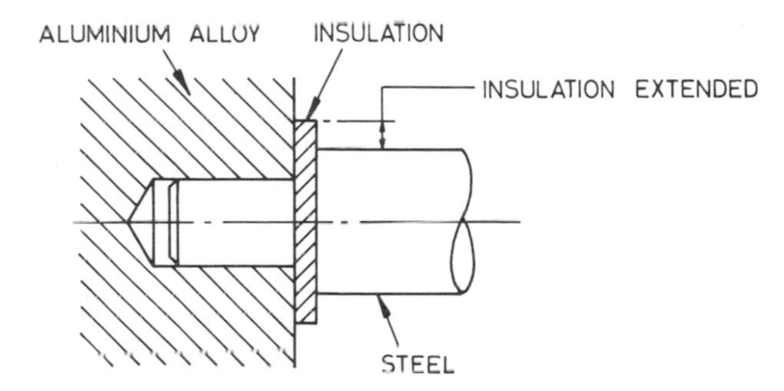

Fig. 9.9 Design to prevent galvanic corrosion

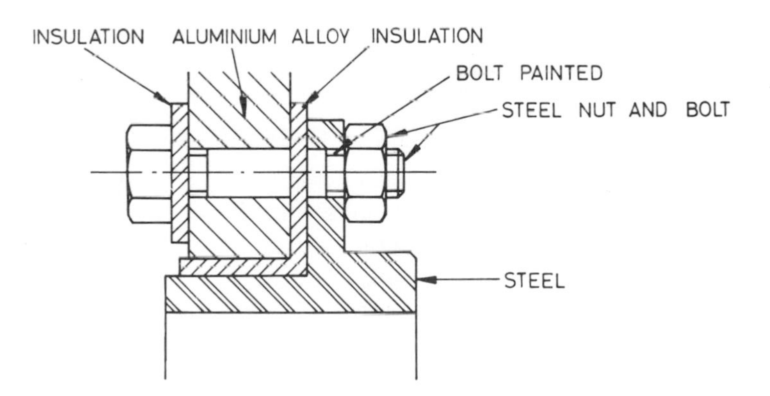

Fig. 9.10 Design to prevent galvanic corrosion

(c) The treatment of the environment

When a confined environment is involved it is often effective to treat it instead of, or in addition to treating the metal itself. For example, air conditioning can be used to remove atmospheric pollution or to reduce humidity; volatile corrosion inhibitors can be introduced in sealed packages (as when photographic equipment is exported) and inhibitors can be dissolved in water or in aqueous solutions (such as motor car anti-freeze solution).

(d) The surface-treatment of metals

Metal can often be adequately protected by covering it with oil or grease when not in use, or by coating it with a thick bituminous type of material that is forced into regions where water may enter or accumulate. (The anti-corrosion treatment of a motor car body is a typical example of the latter.)

Other methods of treating the metal can be classified as:

- oxide coating;
- chemical treatment;
- organic coating;
- metallic coating.

In the following sections, these methods are studied with respect to specific metals, but in all cases it is essential that the metal be properly prepared before it is treated.

When metal is to be painted, a simple cleaning operation is sufficient—such as wiping large surfaces with paraffin or degreasing small parts in baskets over boiling trichlorethene. Larger parts can be immersed in a trichlorethene bath or in an alkaline solution such as caustic soda. (The latter is inexpensive but rather unpleasant to use.)

Parts that have been heat-treated or fabricated using heat must be shot-blasted or pickled in an acid bath to ensure that they are freed from scale. It is essential that metal to be electroplated is particularly clean and is scale-free.

9.3 Corrosion-resisting steels and protective treatments applied to steels

Although steel rapidly corrodes it does not always need to be given special treatment. For example, steel gears that run in an oil bath to lubricate them are adequately protected by the oil.

A number of corrosion-resisting steels are available, but these have their limitations. The ferritic stainless steels (alloy steels with a high chromium and a very low carbon content) and the austenitic stainless steels (alloy steels with a high chromium and high nickel content) can be work-hardened but not strengthened by heat-treatment. The martensitic stainless steels (alloy steels with a high chromium content and about 0.3% carbon) can be strengthened by heat-treatment but are unsuitable for joining by welding because they are air-hardening, and become hard and brittle as a result of the heating and cooling during welding. Because of these limitations, and because of their high cost, stainless steels are not always used in conditions where corrosion is to be resisted. As an alternative other steels are used and given suitable protective treatment.

The protective treatments can be *direct* or *sacrificial.* Direct protection is obtained by coating the surface to keep moisture or corrosive medium away from it; such treatment is only effective as long as the coverage is complete, and this depends upon the application of the protective coating and its resistance to attack. Sacrificial protection is obtained by coating the surface with a metal that is anodic in the environment so that when moisture is present it is the sacrificial metal that will corrode instead of the metal to be protected; this form of protection does not require a full coverage of the surface, but it is always aimed at because up to the time when the surface is uncovered by damage or corrosion of the protective coating, the protection is direct. Oxide coatings, organic coatings and certain metallic coatings (such as tin and nickel) give direct protection to steel, and certain metallic coatings (such as zinc and cadmium) give it sacrificial protection.

(a) Oxide coating

A direct protection is obtained by heating steel to produce an oxide coating on its surface. The colour of the oxide depends upon the temperature to which the steel is heated. The blue finish obtained by heating to 300° C is a popular finish for bolts, springs and small tools. It is hard and not easily chipped but is thin and provides only a small degree of protection; articles so treated are usually oiled or greased for further protection. A heavier oxide is produced by the Bower–Barff process in which steel is heated alternately in air and superheated steam at about 800° C. An attractive black finish is given to small parts by immersing them in a caustic alkali solution containing an oxidising agent (the solution being heated to 140° C before the parts are immersed in it). This treatment gives only slight protection and is unsuitable for assemblies that are of riveted or similar construction because the salts tend to be trapped between the parts.

Vitreous enamel, which is basically borosilicate glass, is used to protect and decorate cookers and other kitchen equipment. It gives a very hard and durable finish that can be easily kept clean. It is supplied in the form of small particles (called *frits*) that are suspended in water into which the pickled steel is dipped, so that when it is removed it will be covered with the particles. After drying, the steel is fired to cause the glass to fuse and coat the surface. The process can be repeated several times to obtain a full coverage or a thicker coating and a coloured surface can be obtained by adding oxides when the frits are manufactured. The fusion temperature depends upon the metallic oxides.

(b) Chemical treatment

The most important chemical treatment is phosphating. This treatment does not, in itself, give a rust-proof coating, but it improves the adhesion of paint, oil or lacquer that must then be applied to the metal. It also minimises the lateral spread of corrosion below the paint or other coating. Phosphating lends itself to the automatic treatment of motor-car body parts, cycle parts and refrigerators prior to enamelling. Zinc-plated and galvanised steel can be phosphated to improve the poor adhesion of paint to zinc. Typical phosphating systems are Coslettising (using an iron phosphate solution), Parkerising (using a manganese phosphate solution) and Bonderising or Granodising (using a zinc phosphate solution).

(c) Organic coatings

Organic coating is the most widely used protective system. Paint has become more useful for protective and decorative systems since synthetic resins have replaced natural resins, making the mixing easier and reducing the drying time. Cellulose is rapid-drying and produces a hard, water-resistant, high-gloss finish which can be polished to a high lustre, which is ideal for motor-car bodywork.

Bituminous paints give good service under normal conditions but tend to become softer when heated and to craze when cold. (The latter effect can be reduced by using a primer, such as red lead.) Stoving japons, based upon bitumen and baked at a high temperature, give a hard durable gloss finish at low cost; they are used for hardware, locks and bolts.

A plastic coating can be applied both for protection and decoration. A typical method is the *fluidised-bed* system in which fine plastics powder is fluidised in a tank by a rising current of air that enters the tank through a porous plate at the bottom. The metal to be coated is heated before it is immersed in the fluidised plastics with which it becomes uniformly coated; the coating is subsequently fused to give a uniform layer. This method is used to coat wire and garden furniture.

(d) Metallic coating

A metallic coating can, as already stated, be applied to give direct protection or give sacrificial protection. The method used to apply the coating depends upon the size of the part to be protected, the degree of protection that is required, the permissible cost of the treatment and the required appearance of the product.

(i) Spraying. Metals can be sprayed on to the surface using a special gun that includes a heating system and is fed with metal in wire or powder form. Large structures, such as bridges, are often sprayed with zinc or aluminium as a primer before being painted. The metal must be roughened by, for example, shot-blasting before being sprayed with metal because its adhesion is purely mechanical. A good coverage is obtained because the particles

are spread and flattened when they strike the surface. To obtain a degree of protection equal to that of a metallic coating applied by hot-dipping or electroplating, the thickness of a sprayed coating must be considerably greater to compensate for oxidation and the formation of cavities.

(ii) Vacuum Deposition. A metal coating can be applied using vacuum deposition by placing the parts to be treated in a jig around the metal to be deposited, within the deposition chamber. The metal to be deposited is wound around a tungsten filament through which a heavy current is passed to vaporise the metal. The chamber must be evacuated to permit the vaporised metal particles to travel, unrestricted, in straight lines with uniform velocity so that when they strike the surface to be coated (which is at a lower temperature) condensation occurs and a thin film is produced. When pressure is restored the parts are removed from the jig and often covered with a coating of lacquer to give added protection.

(iii) Hot-dipping. Galvanising is a hot-dipping process in which rolled sections (steel window frames, buckets etc.) are given a high degree of protection. After cleaning by pickling, a flux is applied and the parts to be treated are dipped in molten zinc (at 450–460° C). When, for example, nuts and bolts are galvanised, they are shaken or centrifuged to remove excess zinc from the threads. Care must be taken when storing galvanised steel because it will deteriorate if stored, without further treatment, in damp conditions. It must be weathered or, preferably, phosphated or chromated for maximum paint-adhesion. Zinc gives steel sacrificial protection.

Tin-plate is produced by passing pickled steel plate through a bath of molten tin at 315–320° C. As it enters the tin the steel passes through a layer of flux on the surface of the tin: as it leaves the bath it passes through a layer of palm oil that prevents the oxidation of the surface of the tin in the bath and improves the protection of the steel because some of the oil adheres to its surface. The protection by the tin is direct, and so the tin should be free from porosity; but pore-free coatings are almost impossible to achieve under commercial conditions. The principal application of tin-plate is for food cans. Corrosion of the steel may take place when the cans contain acid products, and because of the sulphur (added to preserve food) blacking will occur when meat, and some vegetables, are canned unless the tin-plate is lacquered.

A lead coating is obtained by dipping steel into a bath of lead that must include a proportion of tin, antimony or cadmium because steel and lead will not form an alloy. Lead coating is a good primer for painting because paints adhere very well to it. It gives steel direct protection.

(iv) Cladding. This is a diffusion welding process (see chapter 8) in which a sandwich of steel with a corrosion-resisting metal on either side of it is hot-rolled to produce a steel sheet with a thin layer of the protecting metal on each side. Niclad, which is steel clad with nickel, is a typical example. The protection is direct.

(v) Cementation. This is a process whereby the composition of a material is changed by surrounding it with another material in powder form and heating both to a temperature that is lower than that of the melting temperature of both the materials involved. The carburising of steel (which is part of the case-hardening process) is a typical example of cementation. Sherardising (producing a zinc case), chromising (producing a chromium case) and calorising (producing an aluminium case) are typical cementation processes used to protect steel. The steel must be sand-blasted or shot-blasted before it is treated because scale or impurities on its surface will prevent cementation from taking place. As in the previous protection methods, the type of protection depends upon the composition of the case that is produced.

(vi) Electroplating. Metallic coatings can be applied in a plating bath in which the part to be plated is the cathode and the plating metal is either introduced into the electrolyte, or forms a reactive anode that is consumed during the process. Nickel-plating is applied as an undercoat for chromium plating and is important for both protection and decoration. The thickness of nickel deposit depends upon the application (at least about 0.03 mm is necessary for outdoor service but 0.02 mm is considered adequate for indoor service). An ordinary nickel deposit is too dull for decorative work and requires an expensive polishing operation, but bright nickel plating, using additives in the bath, produces a more suitable finish. Chromium-plating that is used for decorative purposes is only about 0.0005 mm thick and is highly porous. It adds little to the protection obtained by the nickel plating but it protects the nickel against tarnishing. It is usually polished to fill the pores.

Cadmium-plating and zinc-plating are applied in deposits about 0.010 mm thick. Cadmium-plating is very similar to zinc-plating but it is more attractive, less liable to finger marking and staining, is more readily soldered, and has a longer life than zinc-plating in marine and highly humid atmospheres; but zinc-plating is better in industrial atmospheres. Cadmium oxide is toxic, and therefore cadmium-plating should not be used for articles which are liable to come into contact with foodstuffs or drinking water. Cadmium-plating and zinc-plating can corrode, and their protective value is no better than their resistance to corrosion.

Parts to be plated should be of simple contour to enable the deposit to be of uniform thickness. Surfaces that need to be polished should be as simple as possible: intricate detail is likely to become blurred when polished, sharp edges and corners will become rounded, and recesses will probably become filled with polishing compound. The overall shape is important because the trapping of air bubbles during plating will result in badly-plated, or non-plated areas. Drainage of the solution must be considered and drainage holes may need to be provided. Means of suspending the part in the solution and of making an electrical connection must also be provided.

9.4 Protective treatment of aluminium alloys

(a) Oxide coating

The natural oxide formed on the surface of aluminium alloys can be thickened by anodising. In this procees the part to be protected is made the anode in a suitable electrolyte (the composition of which depends upon the characteristics required of the film produced) and the cathode is a plate of lead or stainless steel. When an electric current is passed, oxygen is produced at the anode and immediately combines with the aluminium to cause a layer of aluminium oxide to grow into and outwards from the surface of the aluminium. Anodising increases the oxide thickness from about 0.00001 mm to about 0.010 mm in about 30 minutes and the anodised surface can be impregnated with organic materials or coloured. The oxide film is porous and should be sealed for maximum corrosion-resistance either by heating or by applying a coating of, for example, linseed oil to it.

Vitreous enamel can be applied but it must have a firing temperature of not more than 500° C. Work-hardening aluminium alloys will be annealed at the firing temperature and become soft, but the firing can be combined with the solution-treatment of the heat-treatable alloys because the quenching that is part of the treatment will not damage the enamel.

(b) Chemical treatment

The natural oxide can be modified by chemical treatment (*conversion coating*) to produce a phosphate, oxide and/or a chromate film prior to priming and painting. This treatment consists of the application, by brush, dip or spray, of a chemical which is removed after it has acted.

(c) Organic coating

Paint is a commonly-used organic coating. It usually follows chemical treatment but must be applied after priming with an oxide or chromate to suit the paint. A coat of clear or coloured lacquer can be applied to aluminium and its alloys to preserve its brightness or for decoration. Stoving finishes can be applied by spraying and stoved at 90–180° C for up to one hour.

(d) Metallic coating

Metal is applied to the surface of aluminium and its alloys for decoration or to produce special properties. Copper, brass or silver can applied to produce a surface that can be joined by normal soldering. Silver produces a surface with high electrical conductivity, chromium produces wear-resistance and brass allows bonding with rubber. As aluminium is anodic to most of the metals that are deposited on it the protection will be direct and so the coating must be free from pores and discontinuities. Aluminium alloys of the 'working' group can be clad with commercially pure aluminium to combine the strength of the alloy with the corrosion-resistance of the commercially pure aluminium (Cladding is a diffusion bonding process—described in section 8.3(c)—in which a sandwich of thick metal and thinner metal on each side is hot-rolled to produce sheet with a thin layer of the protecting metal on each side.)

9.5 Protective treatment of magnesium alloys

Magnesium and its alloys will, like aluminium, acquire an oxide film, but this film is porous and gives sufficient protection only when in dry air. If the atmosphere in service is humid and if it contains traces of salts, a protective treatment must be applied. As shown in tables 9.1 and 9.2, magnesium is anodic to all other engineering metals and so it is particularly important that it be insulated from these metals in assemblies.

The surface of magnesium and its alloys must be cleared from corrosion-promoting impurities before surface treatment is applied. Fluoride anodising has been developed for this purpose. It is a similar process to anodising as applied to aluminium alloys and which, in addition to cleaning the surface, leaves a film. This film does not, in itself, give protection, but it may be used as a base for paint without further surface treatment.

(a) Chemical treatment

For maximum protection, fluoride anodising should be followed by chromating, in which the material is immersed in a bath. There is a range of chromating processes that satisfies various requirements such as dimensional tolerance, cost and limitation of temperature to which the material can be heated during the treatment. The processes use different solutions and treatment temperatures with treatment times as short as a quick dip or as long as two hours immersion. Chromating is a preparation treatment for painting.

(b) Organic coating

Magnesium alloy components can be given a surface sealing treatment in which they are impregnated with a flexible, durable, water-resistant resin to exclude corrosive media. After the fluoride anodising treatment, followed if necessary by chromating, the parts are heated, coated with resin of the Araldite type and then stoved. This treatment is repeated to produce a total of three coats before final stoving to cure the resin. The surface may need to be painted for outdoor service.

Whatever the previous treatment the surface should, before being painted, be primed using a priming paint containing zinc chromate. This may be followed by more than one undercoat after which any finishing paint that is compatible with the priming and undercoat paint can be applied.

10 Plastics Materials

A plastics material can be defined as 'an organic material which at some time in its history is capable of flow and which, upon the application of adequate heat and pressure, can be caused to take up a desired shape which will be retained when the applied heat and pressure is withdrawn'.

Plastics are extensively used in engineering because they are corrosion-resisting, have low electrical and thermal conductivities, help damp out vibrations, have a high strength-to-weight ratio, are self-coloured and are often less expensive than metals and cheaper to manipulate. Some have special properties which make them useful in certain applications: for example, the polyamides (such as nylon) and polytetrafluoroethylene (PTFE) have a very low coefficient of friction with metals and can be made into light-duty bearings that do not require lubrication. Similarly, polymethyl methacrylate in sheet form (Perspex) is used as an alternative to glass for machine guards and windshields because it has almost the same refractive index as glass but it does not shatter. Plastics materials can be made self-coloured and therefore do not require painting. Many plastics materials can be chromium plated for special effects.

It must be realised that plastics materials are weak compared with metals and have a low stiffness and a high coefficient of thermal expansion. They also differ from metals in that their deformation is dependent upon the rate of loading, the duration of the strain and the temperature, and that they deteriorate under certain conditions. Because of their low strength and low softening temperature they can be easily manipulated and this ease of manipulation into complicated shapes is often the reason for their use.

Although a detailed study is outside the scope of this book, a brief description of the grouping of plastics materials and of their manipulation is necessary to relate the design of plastics products to that of metal products that is the subject of the preceding chapters.

10.1 Thermoplastic materials

Plastics materials are composed of long, and often complicated, molecular chains. In the case of thermoplastics (or thermosoftening plastics) these chains are held together mechanically, in a similar way to that in which wool fibres are joined together to form yarn. When these thermoplastic materials are heated the chains move slightly apart and can be made to slide past each other to take up new positions, which are retained when the heat and pressure is removed. They reach the manipulator in the form of sheets, films (thin sheets), rods, tubes and moulding powders or moulding granules. They can be manipulated with very little force at temperatures only slightly higher than that of boiling water, and joining can be done using a suitable solvent or, in some cases, by the application of heat and pressure.

These materials do not melt, but they flow at appropriate temperatures and pressures and are particularly suitable for manipulation by injection-moulding and extrusion-moulding. They behave like glass when they are blown and can be formed into bottle and dome-like shapes by pressure-forming and by vacuum-forming.

10.2 Thermosets and two-part polymers

Plastics of the thermoset group undergo a chemical change when subjected to a *curing* operation in which they are subjected to heat and pressure; two-part polymers undergo a similar change when they are mixed together. The chemical change produces a three-dimensional cross-linking of the molecular chains which prevents them from sliding past each other. As a result of this, the material cannot be further manipulated by the application of heat and pressure and it cannot be joined using a solvent because it is virtually impossible for the solvent to enter between the chains to disperse them.

Thermosets reach the manipulator as moulding powders, resin-impregnated paper or cloth, or resins for casting or for use in a lay-up process. The two-part materials, as implied, are in the form of two liquids or a liquid and a powder.

10.3 Design for the manipulation of plastics materials

The manipulation processes for plastics can be classified as:
(a) moulding;
(b) pressing and forming;
(c) casting;
(d) laminating.

(a) Design for moulding

Moulding is a manipulation process in which plastics material in the form of dough, powder or granules is subjected to heat (so that it becomes a soft mass or a viscous liquid) and pressure (to form it into shape).

Injection-moulding is used for small parts, such as electrical plugs. This process has been developed from, and resembles, the pressure die-casting process used to manipulate metals and can be applied easily to thermoplastics materials. It can only be applied to thermosets when special equipment is used, otherwise the material will be cured before the mould is filled. Design for injection-moulding is similar to that for pressure die-casting of metal; metal inserts can be introduced by 'moulding around'.

Compression-moulding is very often used to manipulate thermosets because there is little possibility of the material being cured before it takes up the shape of the mould since the moulding powder is introduced into it and the upper part of the press set (the *punch*) descends to compress the material and cause it to take up the shape of the mould before heating starts. It is also applied to thermoplastics when the product is too large to allow injection-moulding to be used. There are variations of the compression-moulding process that are used when awkward materials are moulded, and when inserts are to be introduced by 'moulding around'.

In general the design of parts for compression-moulding is similar to that for gravity die-casting. To some extent the problems are also similar to those associated with injection-moulding and pressure die-casting; but as the process is slower, and is less fully automated, parts that require easing from the mould can be produced.

Extrusion-moulding is the equivalent of the extruding process used for metal and the design considerations are similar.

(b) Design for pressing and for forming

Both pressing and forming are used to manipulate sheet plastics by heating, to make them rubber-like, and then applying a force.

The product is the plastics equivalent of medium-size sheet metal products produced by non-cutting presswork.

Pressing involves pushing the sheet using a moving former to control its inside form and produce a shape like a flower pot.

Forming processes use air to blow or to suck the material into a mould which controls its outside form.

Parts to be produced by pressing and by forming must be of uniform wall section. Also they must have a fairly simple shape (so that the sheet can take it up) and be shaped so that they can be removed from the former or the mould. The design considerations are therefore similar to those associated with sheet-metal presswork.

(c) Design for casting

Casting is not commonly used to manipulate plastics because their viscosity cannot be reduced sufficiently by heating to allow them to flow as easily as liquid metals. Certain thermosetting plastics, nylon for example, can be mixed to produce free-flowing liquids that can be cast, and the exothermal reaction that produces the cross-linking gives the moulding effect. Design for the casting of plastics is similar to that for casting of metals by gravity die-casting, but intricate castings with re-entrant features can be produced by using thin-walled rubber moulds that can be peeled away when the material is solid. Unfortunately the rate of production using this method is very low.

(d) Design for laminates

Laminates can be classified as high-pressure laminates and low-pressure laminates. The former are produced using a press to prevent disruption of the laminate caused by the elimination products formed during the curing of certain plastics. When such products are not formed, plastics can be used to produce low-pressure laminates (but a slight pressure may be used to improve the mechanical properties of the laminate).

High-pressure laminates are usually produced using a fibrous material in sheet form, impregnated with a resin. Typical high-pressure laminates are produced from sheets that are assembled, heated and then pressed to form the laminate, or wrapped on a metal mandrel which is then revolved whilst under pressure between two heated rollers to form a tube. (A rod or a cylinder can be produced using a similar system.)

Low-pressure laminates consist of a reinforcing material and either a thermoset or a two-part polymer which, when cured, becomes the binder. The reinforcing material, usually glass-fibre mat or roving (but sometimes as a woven shape), is applied in layers, by hand, to a former or in a mould to produce the required shape. The binder material is applied by brush or spray between each layer and the surface of the impregrated lay-up may be hand-rolled to compact the reinforcement and distribute the binder material. The binder is then cured at between room temperature and 100° C depending upon its composition. The former or mould need only to be strong enough to support the lay-up and can be made of plaster, wood or a light alloy. Depending upon the material from which the former or mould is made, and the curing temperature of the binder material, the curing may take place before the lay-up is removed from the former or mould.

Fairly large parts can be produced by this method, and the quantities need not be large because the capital costs are relatively low—but the rate of production is low. The product is basically of similar shape to that produced by sheet-metal pressing but is less limited because the lay-up can be removed more easily from the former or the mould (which can be made in sections for the purpose). Also the problem of suiting the shape to the direction of press movement does not exist. The success of the process does, however, depend upon the skill of the personnel.

10.4 Fastening and joining of plastics

Plastics components can be fastened using a nut and screw thread system in a similar way to the fastening system used for metal components. A modified screw thread can be produced by moulding and, if coarse enough, may be strong enough. Parts that will not require separation or will only need to be separated very occasionally can be joined by self tapping screws, but those that will need to be separated more frequently usually require metal inserts with stronger screw threads. These inserts may be introduced by 'moulding around' in a similar way to the 'casting around' method used to incorporate inserts in die-cast metal components. Alternatively they may be pressed into a hole produced during moulding. Components can also be secured together by designing them so that they can be sprung into engagement.

Plastics materials can be welded at comparatively low temperatures by softening them and introducing, as a filler, another plastics material (preferably with a lower softening temperature). A process that is similar to the butt-welding of metal can be done by locally softening the surfaces to be joined and butting them together until cool. Thin sheets can be spot-welded or seam-welded by using local high-frequency vibrations to heat them.

10.5 Machining of plastics

Plastics materials can be machined by most of the conventional methods. The suitability of a particular process and the cutting conditions will, as in the case of metals, depend upon the mechanical properties of the material. It must be realised that most plastics have a relatively low softening temperature and will deform if the correct cutting conditions are not maintained. They are often brittle, and so shock loading and vibrations must be prevented. Some plastics are abrasive and tend to dull the cutting edges of the tools and so alter the cutting conditions. Special attention must be paid to the removal of swarf because many of these materials produce a powdery swarf that hinders the cutting action and may even break the tool. These problems may need to be taken into account when selecting the material, or reduced by, for example, designing to minimise distortion during the machining.

11 Ergonomics

Ergonomics can be defined in various ways. Although each reflects the objectives associated with a specific application, they all imply a systematic study of the relationship between people and machines, and between people and their environment when using the machines. The term *machine* implies widely differing devices such as a machine tool, a kitchen sink, a television receiver and an armchair. Whatever the definition, the final objective is always to fit the machine to the person using it, rather than to make the user adapt him or herself to the machine; but as with almost all problems, an ergonomic solution involves compromise.

Fig. 11.1 illustrates the fundamental relationships involved between the user, the machine and the environment. To use the machine, the person must operate it and observe its response to his instructions or moves—this implies communication between the user and the machine. The operator must be able to assume a posture in which he or she can operate the machine with minimum discomfort, taking into account the environment.

Should it be necessary to store the machine when it is not being used, the ease with which it can be stored, prepared for use, transported to the place of use and returned to the place of

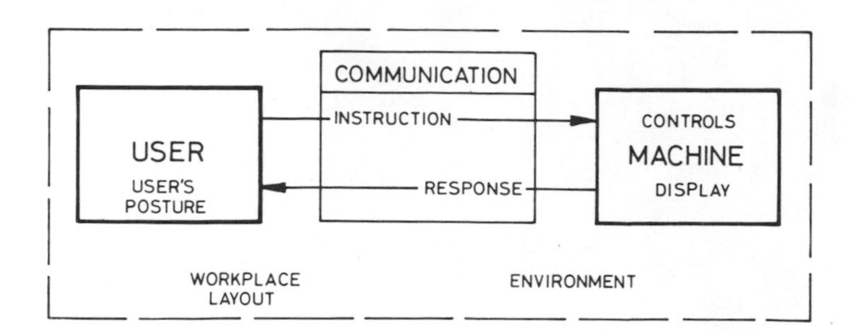

Fig. 11.1 Relationships between user, machine and environment

storage must be considered.

The ergonomic problem areas can be identified as:

(a) the storage, preparation and transportation of the machine;
(b) the environment;
(c) the person's posture when using the machine;
(d) the communication between the user and the machine.

11.1 Storage, preparation and transportation of the machine

The machine should be designed so that the work involved in preparing it for storage and for operation after storage is neither tedious nor unnecessarily awkward. If it must be lifted to take it from the place of storage to the place of operation, then lifting points, positioned so that it is in balance when lifted, should be provided. If carried by the user, its shape must be such that it can be carried safely and without the user's discomfort. If the machine will require routine servicing, it should be designed to provide convenient and safe access to the servicing points. It must also be shaped for easy cleaning.

11.2 The environment

The environment in which a task is performed will affect the efficiency of performance and possibly the health of the operator. The principal environmental factors are lighting, noise and temperature.

(a) Lighting

The amount of light that is required to enable a task to be effectively performed depends mainly upon the degree of detail the operators have to observe, the time allowed for them to see the task, the reflective characteristics of the equipment involved, and the sharpness of their vision. Codes of practice are available that recommend the amount of light necessary for adequate visual performance of various tasks. The immediate background to the task and the general surroundings, such as the walls, ceiling, floor and objects that lie within the operator's vision, must also be considered because his eyes will tend to be attracted to the brighter and more colourful parts of the field of view. The task area should be made the focus of attention by the use of colour and lighting that matches the needs of the task and general level of illumination in the workplace.

Glare often causes discomfort without affecting the operator's ability to see the task, but it may also reduce the visibility. It may be caused by the position of the light sources in the operator's field of vision, their brightness or their area, or be caused by reflection from the working surfaces or from the workpiece itself. Glare can be minimised, or even eliminated, by careful design of the lighting sources and their positions and by changes in the texture of the working surfaces.

Workers will become less tired if the lighting and the colour scheme are arranged so that there is a gradual change in brightness and colour from the task area to the surroundings, and if the work area is located so that the operator can occasionally relax by looking away from the task towards a distant object or surface. The latter should not be so bright that the operator's eyes take time to adjust to the change when he again looks at the task.

(b) Noise

Noise may produce three effects that are not necessarily related: these are annoyance, damage to hearing, and reduction of work efficiency.

The annoying effect of noise varies from individual to individual, but most people find high-pitched noises more annoying than low-pitched noises, and interrupted, or sudden, unexpected noises more annoying than steady ones. Noises that are thought to be unnecessary or due to thoughtlessness, and those of unknown source, are especially irritating. Noise caused by equipment that a person is using is less annoying than that caused by equipment being used by another person because the operator has the option of stopping the noise.

Damage to hearing can be caused by noise that might be regarded as acceptable in ordinary industrial life, but it may take several years for the effects to become bad enough to be serious. The sounds to which the ear becomes especially insensitive tend to be slightly higher in pitch than the noise to which it has been exposed. Experimental studies indicate that people do not work slower when exposed to noise, but it appears that loud noise causes people to be less accurate and liable to make the kind of mistake which produces accidents.

If the level of noise is too high it should be stopped at source by maintenance, silencing and by placing vibrating equipment on isolating mounts to stop the transmission of sound from it. Further protection can be obtained by placing sound-insulating walls around the equipment and by introducing soft, absorbent surfaces in the workplace. Ear plugs or ear muffs should be provided to further reduce the effect of the noise if there is still a hazard.

(c) Temperature

In order to perform a task efficiently the operator should feel neither hot nor cold, but comfortable. The factors that determine how the operator feels are air temperature, radiant temperature, air humidity, rate of air movement, his own response to these conditions, and the particular task that is being performed.

The optimum air temperature for light work in British factories has, by observation, been found to be 18.3° C. When heavy work is done the temperature should be lower, and when office work and other sedentary work is done, it should be higher. When a worker is exposed to radiant heat the air temperature should be lower.

The radiant temperature comfort level is between 16.7 and 20° C and it is important to shield workers from radiant heat when high-temperature processes are used and to avoid excessive loss of heat from the body to cold surfaces such as windows.

Humidity has little effect at ordinary temperatures, but low humidity may cause discomfort through drying of the nose and throat when the temperature is high, and high humidity may cause the sensation of stuffiness in a crowded or ill-ventilated room. Air humidity and air movement become very important at high temperatures because they influence the amount of sweat which can be evaporated from the body surface (a process which reduces the body temperature). When the air and radiant temperatures are correct the ideal rate of air movement is such that it is just perceptible.

11.3 The user's posture

The posture adopted by an operator depends upon the control method, the response display, the layout of the machine or workspace, the seating or other support that is provided, and the shape of the control or handle. These factors must be considered with a view to ensuring that the operator can adopt a posture whereby he can control the machine, or carry out his task safely and efficiently without being unnecessarily fatigued in doing so.

(a) The control method

The choice of control method is considered in section 11.4(a), but it must be realised that the basic posture adopted will depend upon whether the operator controls the machine using his feet, his hands, or both feet and hands, and upon the extent of the controlling movement.

(b) The response display

The nature of the display and its clarity, discussed in section 11.4(b), can be illustrated by the response display used in motor-car controls for the headlights and the road speed. When the driver switches on the headlights at night, he can determine if the lighting is adequate for the conditions whilst looking at the road ahead, and there is no need for any other display except a signal as a warning that the headlights are on full beam. In contrast to this, it is necessary to incorporate a speedometer as a response display to enable the driver to verify his judgement of road speed, and it is necessary for it to be positioned so that it is visible from the driving position.

(c) The layout of the machine or workplace

The layout of the controls and the response display influences the operator's posture and movements. The controls must be positioned so that each can be reached from a normal working position, and so that the user does not need to lean or stoop over the machine or workspace when operating them. They must be arranged and shaped so that each is readily identified. The response displays should be located to allow the operator to judge the effectiveness of his control movements whilst they are being made. Anthropometric charts (giving the average person's measurements) are available to use directly or to make *lay figures* for laying out the controls.

(d) The shape of the control device

The shape of the control device with respect to display and desired mechanical advantage is discussed in section 11.4(b), but its shape affects the operator's posture. This is illustrated by the hacksaw handles shown in figs 11.2 and 11.3. That shown in fig. 11.2 is tiring to hold and it is difficult for the user to control the frame and prevent the blade from being twisted; but that shown in fig. 11.3 is much less tiring to hold and allows the user to have good control over the frame. Similarly, the shape of a pedal will dictate the position of the user's foot, and possibly of his lower leg.

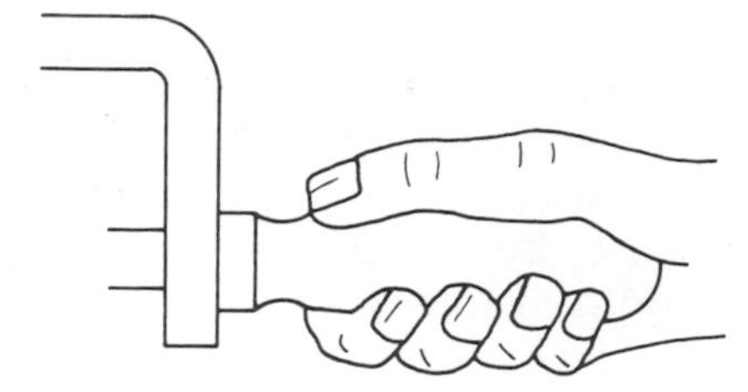

Fig. 11.2 Hacksaw handle that is difficult to hold

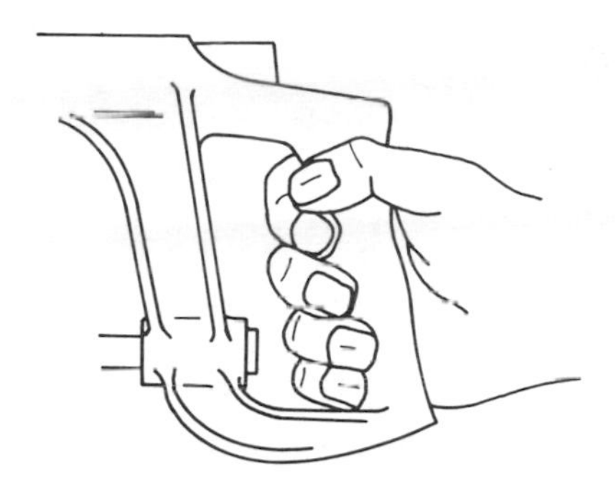

Fig. 11.3 Improved hacksaw handle

(e) Seating and other supports

The Factories Act of 1961, and the Offices, Shops and Railway Premises Act of 1963 both lay down that where work can, or must, be done sitting, a suitable seat should be provided for each person, together with a foot-rest if necessary. The function of the seat is to take the weight off the feet, to support the sitter so that he can maintain a stable posture whilst working, and to relax those muscles which are not required for the work. It should be designed so that there is no discomfort due to pressure on the underside of the thighs, and no restriction of the blood supply to the buttocks because of an unsuitable distribution of the sitter's weight. The seat height should be greater than the length of the lower leg from the floor to the inside of the knee when bent at right angles, and its depth should be less than the distance from the back of the buttocks to the inside of the calves. The seat width must allow space for the hips and lower trunk, and for clothing and contents of the pockets. A range of postures must be catered for because the sitter needs to shift his weight during a spell of work to prevent discomfort such as a 'pins and needles' feeling, and arm rests should be introduced if support points or leverage points for use when getting in and out of the chair are required.

If the worker is to be seated at a table to give support for his work or hands, its height must be related to the height of the seat, and space must be provided beneath it for the worker's thighs and feet.

11.4 Communication between user and machine

The communication system consists of the control device, which may include a graduated dial, and a display or signal to indicate the response of the machine to the instruction.

(a) The type of control device

The type of control device selected depends upon a number of factors: the principal ones are the required speed of operation, the required accuracy of the control and the force that is to be exerted. Figs 11.4 to 11.13 show the more common control devices and table 11.1 shows their suitability with respect to the principal factors. The direction of control device movement to produce *on* or *increase* effect is also shown.

The control device is usually rotated in a clockwise direction or moved to the right for *on* or *increase*, but a different movement is adopted in special circumstances. For example, people expect a downwards movement of an electric light or power switch for *on*. Similarly they instinctively pull a hand-operated lever (such as a motor car hand-brake) towards them for positive action, but push a foot-operated lever away from them for the same effect, because in each case they make the action that enables them to exert maximum power.

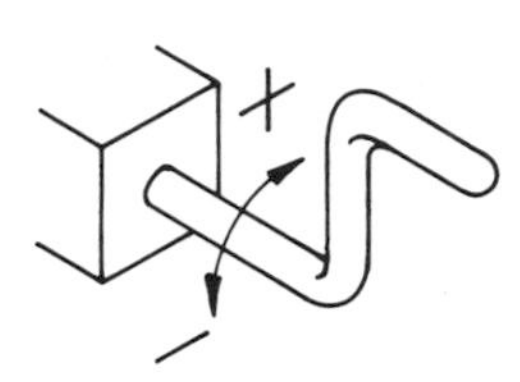

Fig. 11.4 Crank

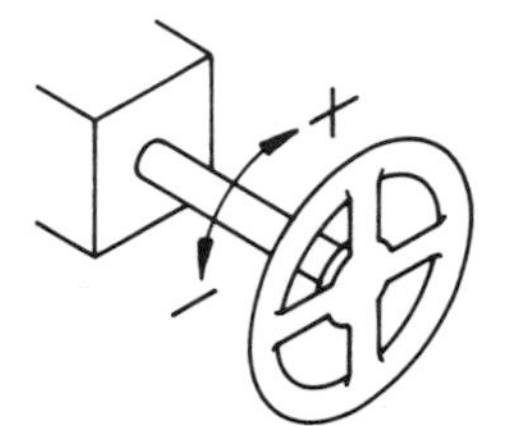

Fig. 11.5 Handwheel

Table 11.1

Control device	Speed of operation	Accuracy of operation	Force exerted	Direction of movement for *on* or *increase*
Small crank (fig. 11.4)	Good	Poor	Unsuitable	Clockwise
Large crank (fig. 11.4)	Poor	Unsuitable	Good	Clockwise
Handwheel (fig. 11.5)	Poor	Good	Fair to poor	Clockwise
Knob (fig. 11.6)	Unsuitable	Fair	Unsuitable	Clockwise
Rotary selector switch (fig. 11.7)	Good	Good	Unsuitable	Clockwise sequence 1–2–3 . . .
Horizontal lever (up and down) (fig. 11.8)	Good	Poor	Poor	Up
Vertical lever (across user's body) (fig. 11.9)	Fair	Fair	Fair	To right
Vertical lever (to and from user's body) (fig. 11.10)	Good	Fair	Good—if lever is long.	Towards user
Push button (fig. 11.11)	Good	Unsuitable	Unsuitable	—
Foot pedal (fig. 11.12)	Good	Poor	Good	—
Joystick selector switch (fig. 11.13)	Good	Good	Poor	—

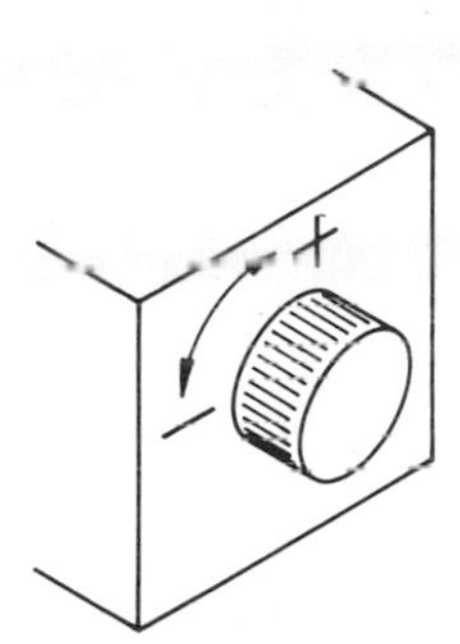

Fig. 11.6 Knob

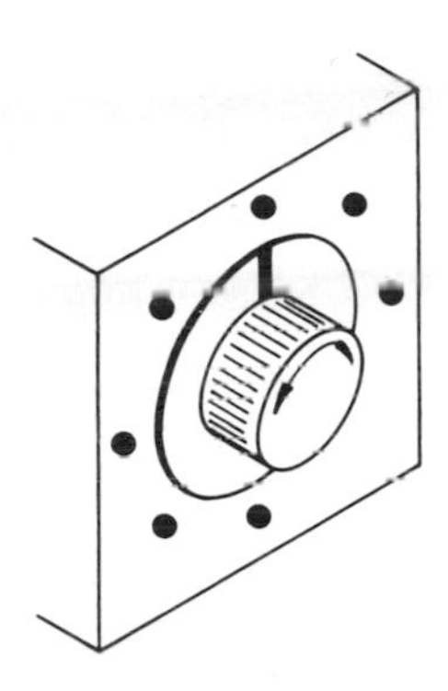

Fig. 11.7 Rotary selector switch

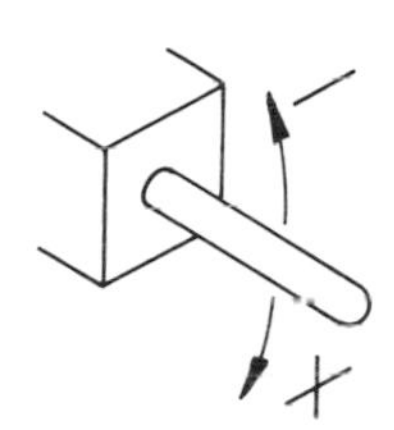

Fig. 11.8 Horizontal lever

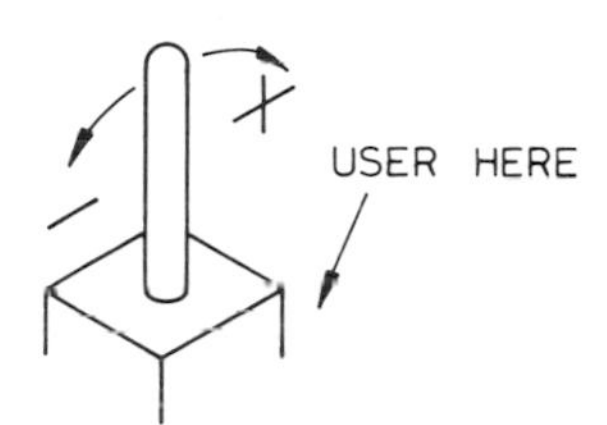

Fig. 11.9 Vertical lever (left and right)

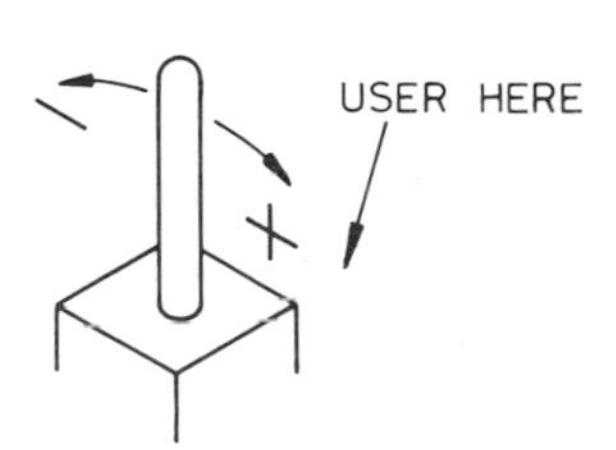

Fig. 11.10 Vertical lever (to and fro)

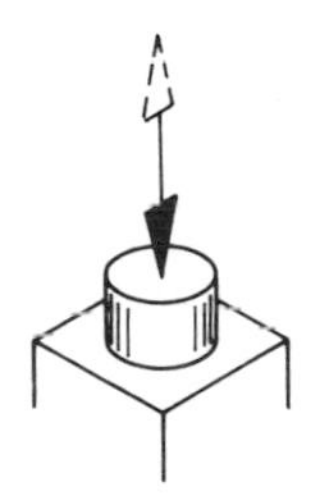

Fig. 11.11 Push button

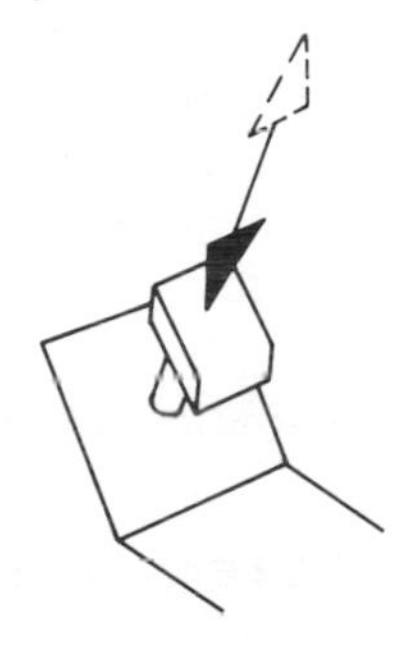

Fig. 11.12 Foot pedal

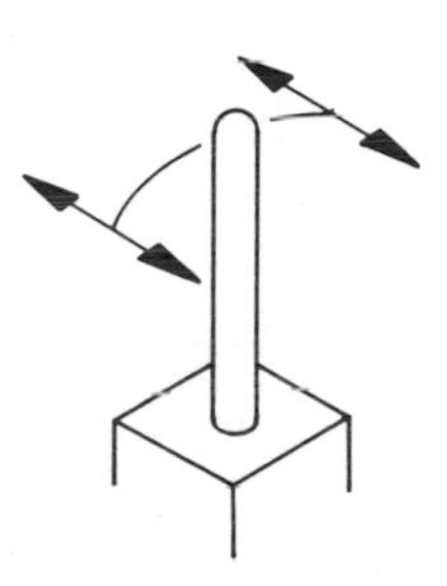

Fig. 11.13 Joystick selector switch

(b) The control layout and the shape of the controls

When the machine is controlled by the sequential operation of a number of control devices they should, if possible, be arranged in that sequence. It is also convenient for the user if related controls are near each other. For example, where motor-car controls are mounted on the steering column, those that are related to warning or signalling to others (driving lights, direction indicators and horn) should be grouped together, and those that are for the driver's convenience (wipers and windscreen water jets) should be grouped together and operated by the other hand.

For ease of identification, the arrangement of a group of controls should be the same as the devices controlled by them. For example, the four controls that operate the four electric cooker hot plates are not readily identified when arranged as in fig. 11.14, by are readily identified when arranged as in fig. 11.15.

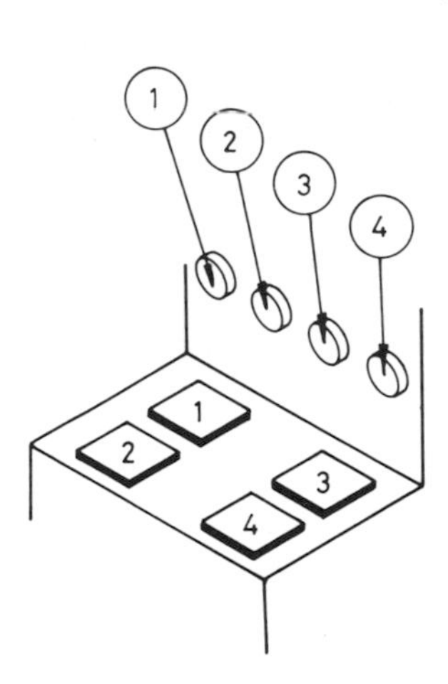

Fig. 11.14 Un-related control arrangement

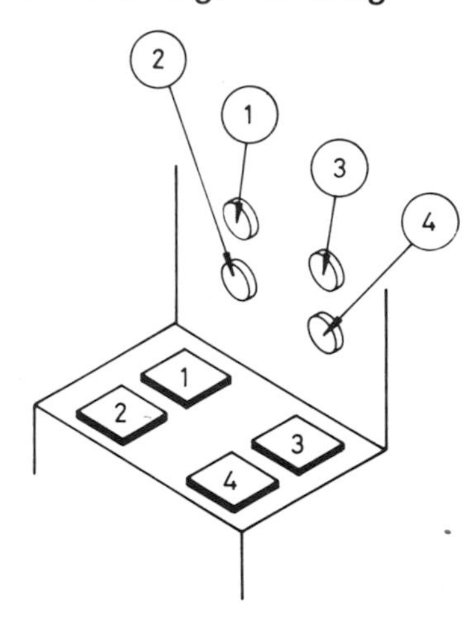

Fig. 11.15 Related control arrangement

When the control device is to be rotated it should be shaped so that its 'pointer end' is obvious; compare the control shapes in figs 11.16 and 11.17.

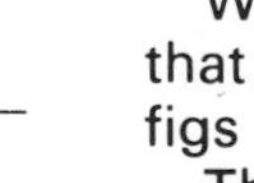

Fig. 11.16 Badly-designed pointer control

The shape and size of the control device should be such that the user is encouraged to handle it in such a way as to exert the required force or control, but not excessive force, damaging the control or the machine. For example, the knob of a rotary switch for a television receiver should be small, so that the user will operate it using just the ends of a thumb and first finger, and not by gripping it to use wrist action.

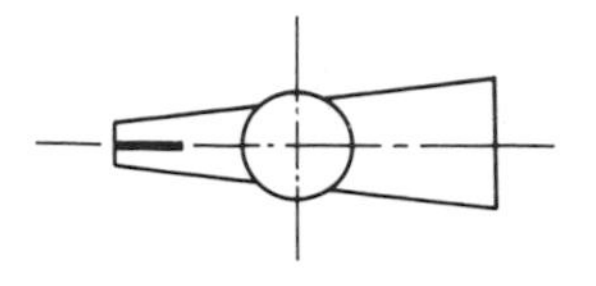

Fig. 11.17 Improved design for a pointer control

(c) The type of response display or signal

The response of the machine to the control may be indicated by a display or signal that is qualitative or by one that is quantative.

A *qualitative display* or signal is used to indicate an 'on' or 'off' response such as valve open or valve closed, power on or power off; it can be visual or be auditory. The visual display may be simply an 'on' or 'off' light, a set of different-coloured lights (as in traffic signals), a light of varying brightness to indicate changing conditions, a set of lights that come on in different combinations for on and off, or a simple flashing light. Auditory indicators cannot convey detailed information unless they actually transmit speech, but they attract immediate attention from any direction and are usually used for warning indication.

A *quantative display* is used when the user requires numerical information from the instrument; the information can be given as a digital display or as an analogue display. A digital display presents the information directly as a number and an analogue display does so by means of a scale and a pointer whose position relative to the scale is analogous to the value it represents. For example, a digital watch indicates the time directly as a number, and an analogue watch indicates, by the hands and the face, the number of times the spindle that carries the minute hand has rotated since the time when the watch was set—which is analogous to the actual time.

A digital display is used where precise readings are required and an analogue display is used where an easily interpreted indication of rate and direction of change is required, or where a quick check of several values will be made.

(d) The relationship between control movement and analogue display

The relationship between the control movement and its effect has already been considered. It is also important that the relationship between the control movement and that of the pointer over the scale is as expected by the user, to reduce the probability of errors. Fig. 11.18 shows the pointer movements associated with a clockwise (positive) movement of a rotary control; the relationship between pointer movement and positive movement of a lever is the same.

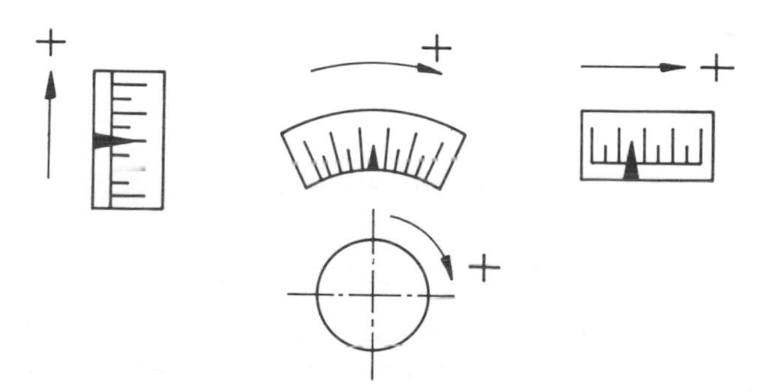

Fig. 11.18 Relationship between pointer movement and that of a rotary control

(e) The design of analogue displays

The display may be in the form of a pointer that moves over a straight scale, one that swings over a curved scale, or one that rotates over a circular scale. The latter enables a long scale to be used in a small space, and is easily read by the user without changing position in order to see the extremities of the scale.

The scale and the dial face should be designed with the minimum amount of markings, and the pointer should be simple and clear. The dial shown in fig. 11.19 is confusing, whereas that shown in fig. 11.20 is clear. When the numbers on the dial are the same as those on a clock, the dial is easier to read when they are in the positions they occupy when on a clock face; compare figs 11.21 and 11.22.

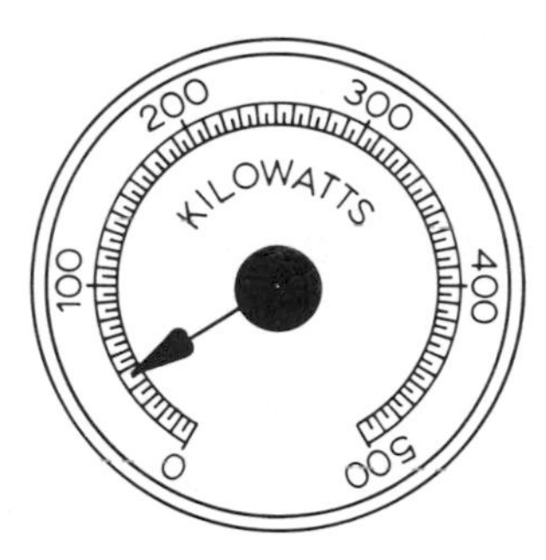

Fig. 11.19 Confusing dial and pointer

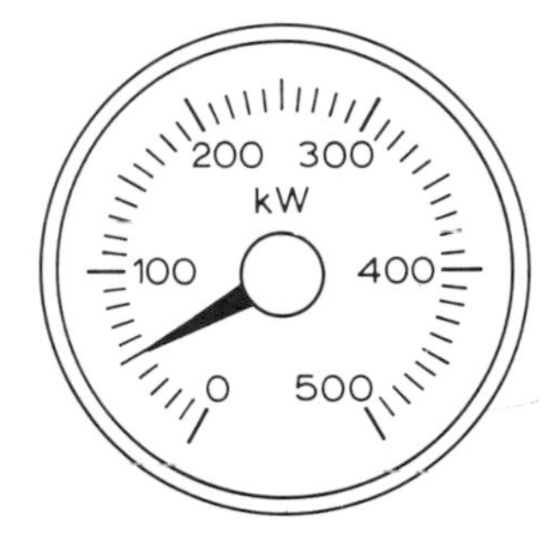

Fig. 11.20 Clear dial and pointer

Fig. 11.21 Numerals not related to clock numerals

Fig. 11.22 Numerals related to clock numerals

When a control and its associated display can be placed near each other, the control device should be placed below, or to the right of, the display, as shown in figs 11.23 and 11.24, so that the user's hand is less likely to interfere with reading the display. Most controls can usually be operated equally well by either hand, leaving the user free to write with the other one. The preferred hand is better for making fine, accurate adjustments, and those controls should be placed in a central position to cater for both right-handed and left-handed users. Labels should be placed above the controls, so that they are always visible, and, to ensure consistency, always over their respective displays as shown in fig. 11.25.

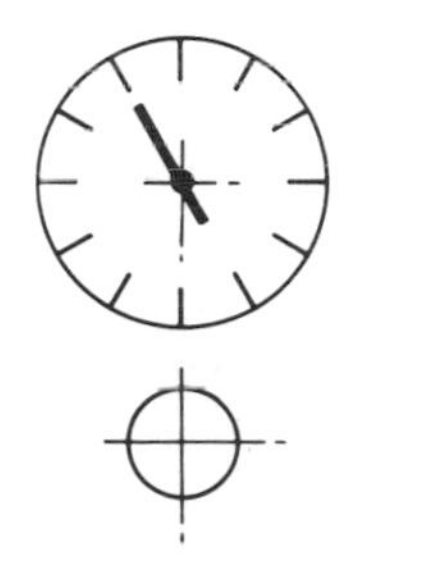

Fig. 11.23 Arrangement for easy reading of display

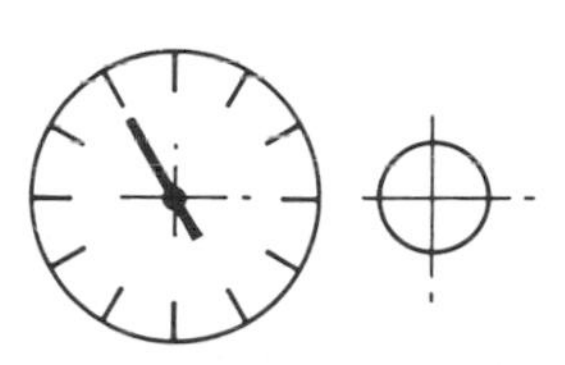

Fig. 11.24 Arrangement for easy reading of display

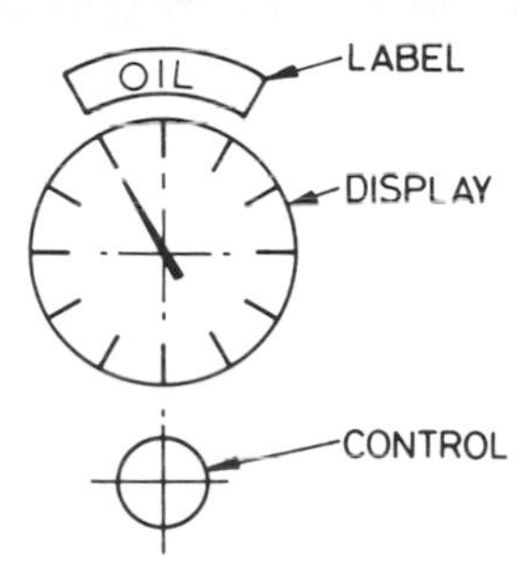

Fig. 11.25 Label, display and control arrangement

(f) The layout of the display panel

The dials on a display panel should be easily identified and read by the user. This is most easily achieved if the dials contrast well with the panel face and are arranged in irregular groups. The arrangement shown in fig. 11.26 makes difficult the identification of each dial, but by arranging the dials as shown in fig. 11.27 each one can be easily identified.

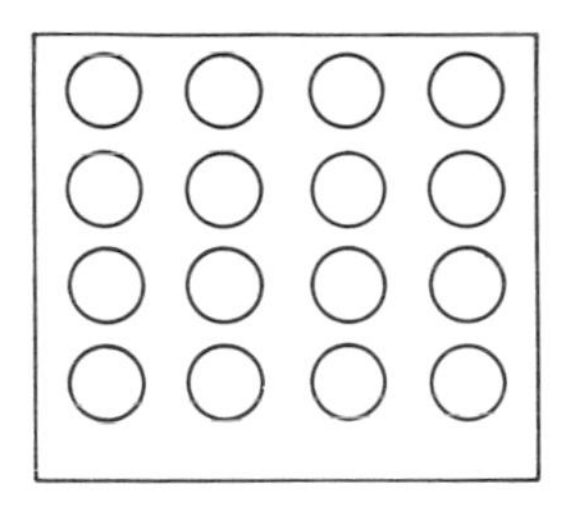

Fig. 11.26 Regular arrangement of dials making identification difficult

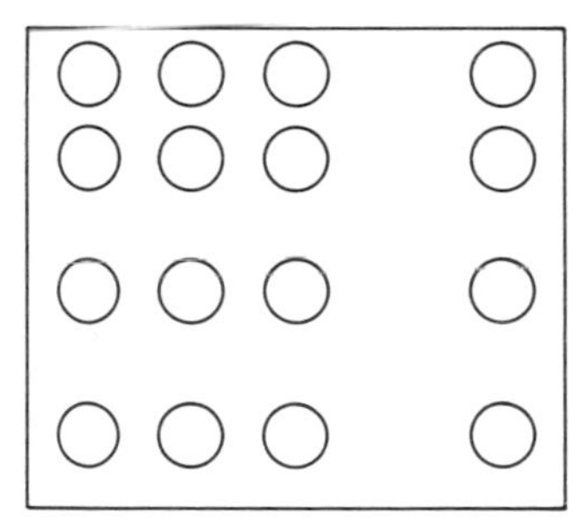

Fig. 11.27 Irregular arrangement of dials making identification easy

The system can be checked for 'normal' or 'all systems correct' if this condition is indicated by all the dial pointers being in the same position; compare the pointer positions figs 11.28 and 11.29.

Fig. 11.28 'Normal' pointer positions haphazard

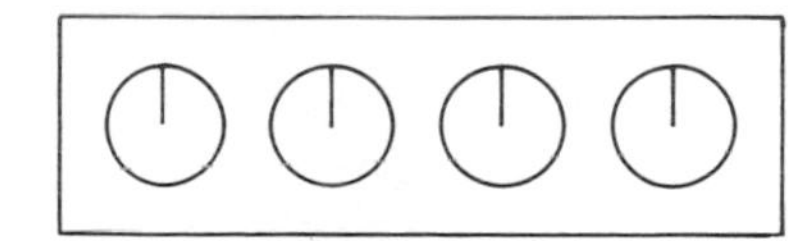

Fig. 11.29 'Normal' pointer positions regular

11.5 Health and safety

The study of ergonomics helps to eliminate irritating things such as noise, because when people are irritated their mental health is affected, and they tend also to make mistakes which may adversely affect their safety and that of others. It also aims at improving the posture of people when at leisure and when working because bad posture can damage their physical health, and because safety depends upon effective control which, in turn is related to correct posture.

The arrangement of controls and displays affects the ability of the user to use and control machines, and it therefore affects both his or her health and safety, and the safety of others.

In addition to fulfilling those tasks, ergonomics enables people to enjoy using their equipment both in work and play.

12 Machine Elements

Machines include basic elements such as shafts, bearings, gears and cams. This chapter describes the application of the more commonly-used elements and their main features. The reader should consult a machine-drawing text book and manufacturers' literature for more details, and a mechanical-science text book for an analysis of the velocity and acceleration of the component parts and of the forces to which they are subjected.

12.1 Transmission of rotational motion using a shaft

The fundamental arrangement shown in fig. 12.1 appears, in one form or another, in all systems in which rotational motion is transmitted.

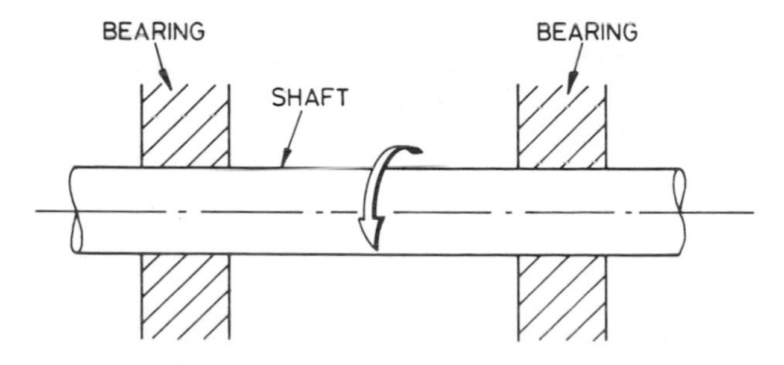

Fig. 12.1 Basic shaft arrangement

(a) Bearings

As the shaft needs to be constrained and supported, the arrangement must include bearings that are designed so that friction does not cause loss of power, or wear that would reduce the accuracy of shaft constraint. Friction can be reduced by using a bearing material with a low friction coefficient, by effective bearing design and by lubricating the bearing surfaces. The introduction of a lubricant presents problems when the bearing is near food (in a food mixer), near washing (in a washing machine or dryer) or when the mechanism is part of a clutch or a similar device that uses friction drive. The forces against which the rotating shaft must be held may act in a radial direction, in an axial direction (thrust), or in both directions, depending upon the system of which the shaft is part. A plain-bearing, which is basically a tube, and therefore able to hold the shaft against radial forces, can, by introducing a collar on the shaft to bear on one of the end faces, also resist thrust, but ball-bearings and roller-bearings are usually designed for best performance in one system of loading only.

Plain-bearings made of a low friction material, such as Nylon, are suitable for light-duty service, and after a short running-in period with a lubricant such as light oil or grease do not usually require further lubrication. They are therefore particularly useful in mechanisms in which a lubricant cannot be applied. When a stronger, and self-lubricating plain-bearing is required, one produced by sintering and then impregnating with oil can be used. Other plain-bearings require a lubrication system.

Ball-bearings are especially suitable for high speed running because they are designed to have *point contact* between each ball and each of the two races. Radial types and thrust types differ in the arrangement of the balls and the races as illustrated by figs 12.2 and 12.3. Radial ball-bearings have a moderate to good radial capacity, and a good to very poor axial capacity (depending upon the specific design). Thrust ball-bearings have a good to very good axial capacity and little or no radial capacity (again depending upon the specific design).

Roller-bearings are designed to have *line contact* between each roller and each of the two races, and can, therefore, carry higher loads than can ball-bearings, but are more suitable for

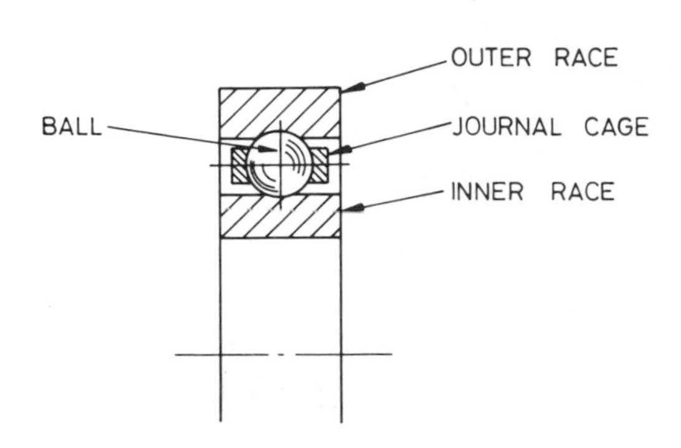

Fig. 12.2 Radial bearing

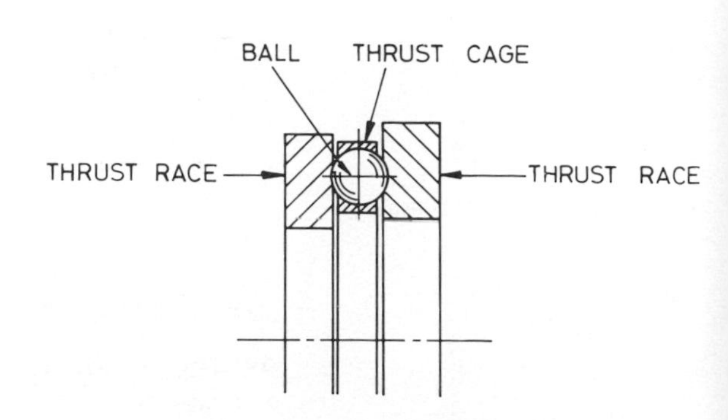

Fig. 12.3 Thrust bearing

running at lower speeds; they also are classified as radial types and thrust types, with similar characteristics to their ball-bearing equivalents. Needle roller-bearings are similar to roller-bearings but they have long, small-diameter rollers (called *needles*) that, in some types, are not held in cages; they have a small outside diameter and can be used in confined spaces, but are only suitable for very light duty, and have no axial capacity.

Although, in theory, there is only point contact in ball-bearings and line contact in roller-bearings, and only rolling motion, it is necessary to introduce a lubricant because in practice there is *area* contact between the balls or rollers and the races, and a sliding motion between them and the cage which retains them. The lubricant, which should be a good-quality mineral oil or grease, also protects the bearing components from corrosion.

(b) Seals

A sealing system must be incorporated when the bearing lubricant might otherwise escape, or when the shaft is part of a gearbox that contains an oil bath. The provision of a seal for a rotating part is more difficult than the provision of a static seal. The particular type of rotating-shaft seal that is used depends upon the diameter of the shaft and the speed of its rotation, the pressure variation across the seal, and the properties and temperature of the fluid to be retained. The usual sealing systems used to retain lubricating oils are compression packing glands and lip seals. In both of these systems the shaft is surrounded by the seal, which must be radially deformed to prevent leakage along the shaft. In the compression packing gland, shown in fig. 12.4, the packing is compressed by the axial force exerted by the tightening system and its continued efficiency depends upon regular maintenance. A lip seal (see fig. 12.5) is used where there is little or no provision for adjustment whilst in service. It is easy to install and its housing is about one quarter of the size of that for a comparable packed gland, but its satisfactory functioning depends to a large extent upon the surface finish of the shaft.

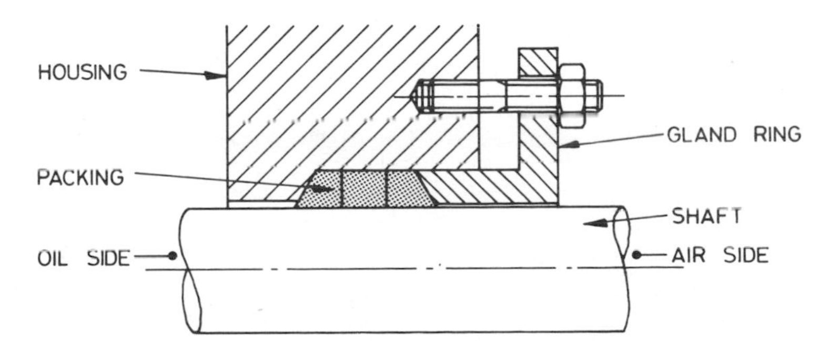

Fig. 12.4 Compression packing gland

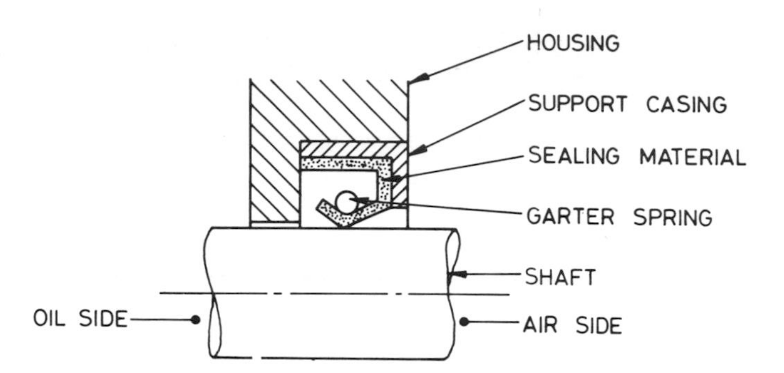

Fig. 12.5 Lip seal

(c) Escapement mechanism

An escapement mechanism is included when a shaft with a constant torque applied to it is to have an intermittent rotational movement. Fig. 12.6 illustrates a typical clock escapement. The escape wheel is mounted on the shaft that is to have the intermittent movement, or on one of the shafts that drive it, and the pallet is mounted on a pivot shaft so that it can swing. In the illustration, the pallet is swinging in an anti-clockwise direction and the pad A is just about to halt the rotational movement of the escape wheel by engaging tooth X. This engagement will cause a recoil action, so that the pallet will swing in a clockwise direction to release tooth X and allow the escape wheel to rotate until pad B engages tooth Y. This engagement will again cause a recoil action, reversing the rotation of the pallet and releasing tooth Y to complete the cycle. The cycle time is adjusted by moving the pendulum bob to change the periodic time of the pendulum. The balance wheel and hair spring of a watch mechanism are together equivalent to the pendulum in this mechanism.

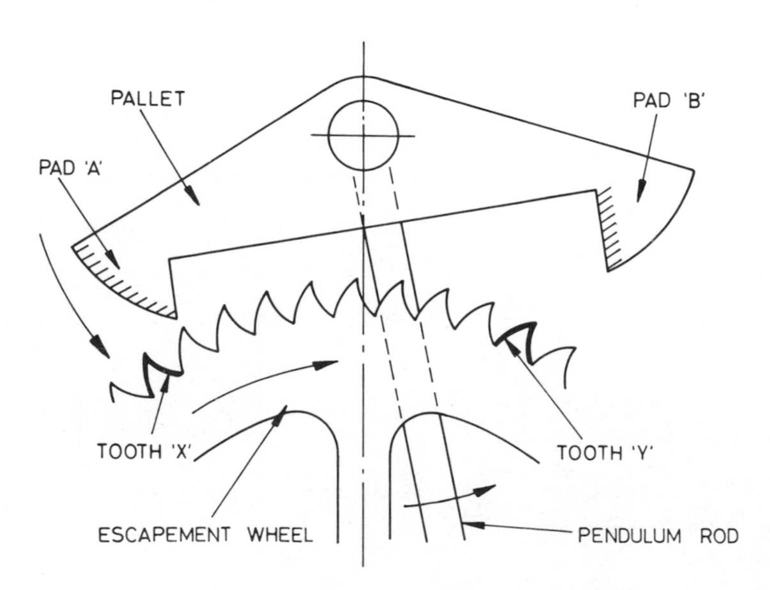

Fig. 12.6 Escapement mechanism

12.2 Transmission by shafts connected end-to-end

Shafts are connected end-to-end when a single shaft is difficult to manufacture or to install, when the component parts to be connected are not in a fixed position, or are not in line, or when the driven component must be disconnected from the driver component.

(a) Couplings

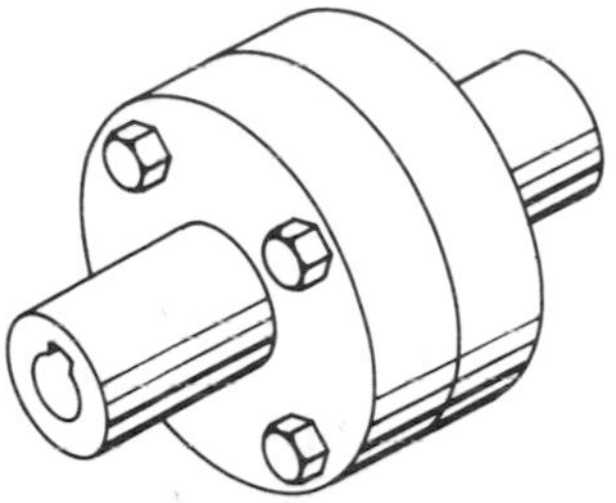

Fig. 12.7 Flanged coupling

A coupling is used to connect two shafts together, end-to-end. It may be rigid, and be unable to compensate for badly aligned shafts, or be flexible so that it can compensate for mis-alignment.

The principal types of rigid coupling are the flanged coupling and the muff coupling. The flanged coupling, shown in fig. 12.7, consists of two flanged components, one keyed to each shaft, and connected together by nuts and bolts that pass through holes in the flanges. The muff coupling (fig. 12.8) is basically a tube into which the ends of the shafts are fitted and secured by keys and setscrews. A variation of the muff coupling, called a split-muff coupling, is used to connect and disconnect shafts when they are in position; it is a muff coupling split along its length so that the two halves can be assembled over the two shafts and secured together by nuts and bolts and so grip the two shafts.

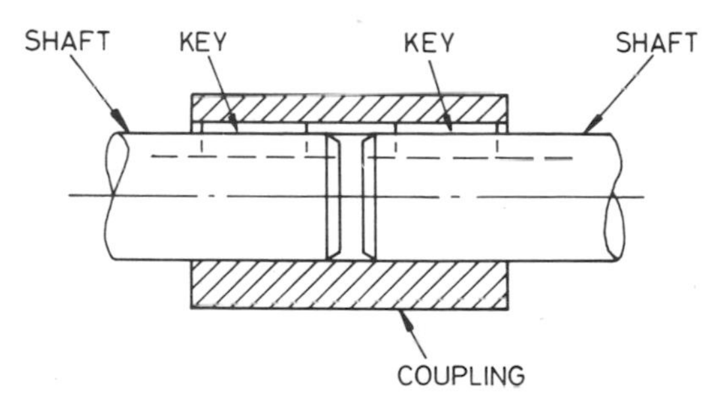

Fig. 12.8 Muff coupling

It should be noted at this point that a key is a piece of metal inserted between a shaft and a part placed on it to secure them together. The key is usually of rectangular section and fits into a keyway, that is also of rectangular section, in the shaft and in the bore of the mating part. (A keyway can be seen in the bore of the coupling member in fig 12.7.) A spline system (see fig. 12.9) is used when a shaft-mounted component must rotate with the shaft and also be able to slide along it. Serrations (see fig. 12.10) are used to secure a component to a shaft.

A flexible coupling is necessary when the shafts cannot be perfectly aligned. These couplings can be made entirely of metal,

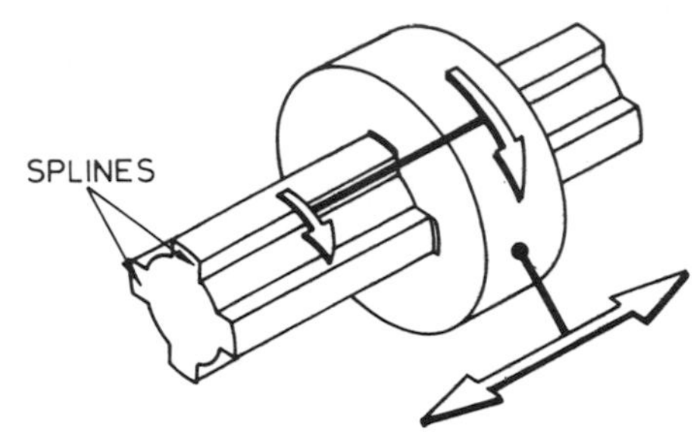

Fig. 12.9 Spline system

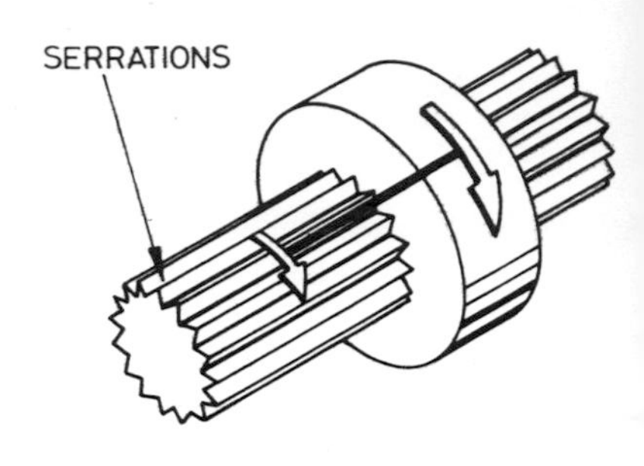

Fig. 12.10 Serration system

or they can made of metal and an organic material. Those that are made entirely of metal are of two main types. One type uses metal springs to obtain the flexibility (the coil-spring coupling shown in fig. 12.11 is a simple example of this type) and the other employs a sliding action between its component parts (the Oldham coupling in fig. 12.12 is typical of the this type). Couplings that include an organic material use rubber, or fabric reinforced with rubber, as the flexible material; this may be in the form of a thick hose instead of the coil spring in the coupling shown in fig. 12.11, or be a thick disc that is sandwiched between the flanges of a flanged coupling and bonded in place.

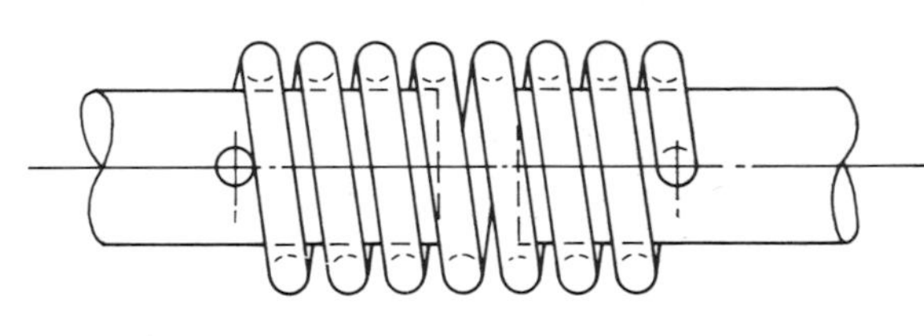

Fig. 12.11 Coil-spring coupling

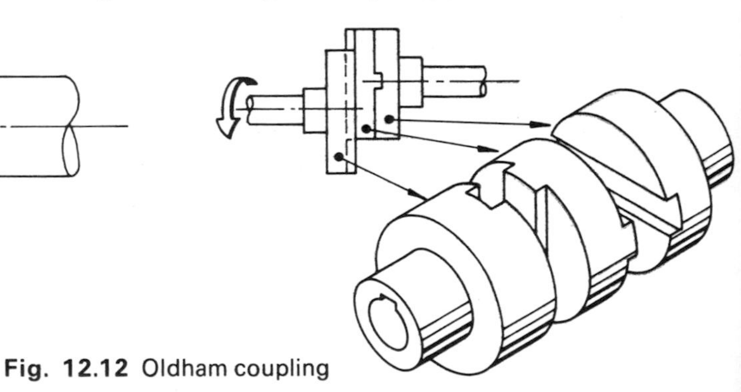

Fig. 12.12 Oldham coupling

(b) Universal joints

A universal joint is required when the axes of the two shafts are inclined at a small angle. A Hooke's coupling (fig 12.13 and 12.14) enables the angle of inclination to change whilst the shafts are rotating. (A typical application is the transmission system for a car with the engine at the front driving the wheels at the rear.) Two couplings of the Hooke type can be used, as shown in fig. 12.15, to connect two shafts whose axes are some way apart if the intermediate shaft can be made long enough to keep the angle of inclination small.

When shafts connected by a Hooke's coupling are inclined, the speed of the output shaft fluctuates—the extent of this fluctu-

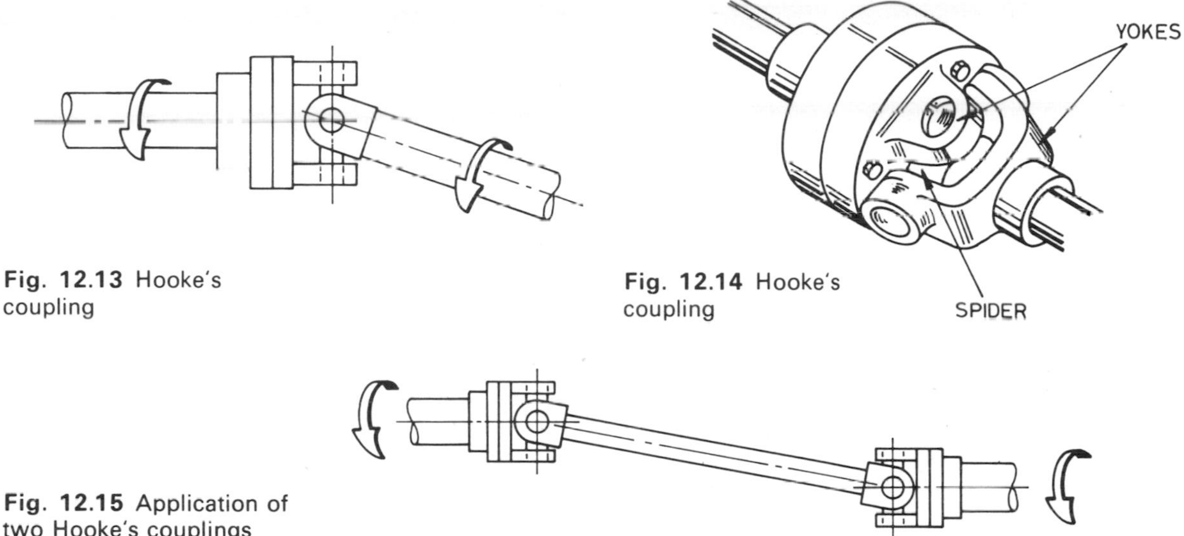

Fig. 12.13 Hooke's coupling

Fig. 12.14 Hooke's coupling

Fig. 12.15 Application of two Hooke's couplings

ation increases with the angle of inclination. To overcome this problem a number of constant-velocity joints have been developed. The Birfield constant-velocity joint comprises a solid spherical component and a hollow spherical component, each of which has six grooves machined in it in line with the shaft axis. The solid component fits inside the hollow component, and is separated from it by a cage containing six balls, each of which engages in a groove in each of the two spherical components to transmit the drive. The balls move in the grooves to compensate for the inclination of the shafts and to produce a constant velocity output.

(c) Clutches

A clutch, which is a form of coupling, is introduced into a system when the two shafts must be engaged or disengaged easily and rapidly to enable the driving shaft to continue to rotate without driving the output shaft when, for example, a gear-change is made. A disc clutch consists, in principle, of two flanges with a disc of a high friction coefficient material riveted to one of them; in the engaged position the two flanges are held together by springs so that the drive is transmitted by friction. A cone clutch is similar, except that the engaging surfaces are conical and the friction material is secured to the male cone. A type of clutch called a claw coupling is used when the two shafts need to be engaged for only a short time. Fig. 12.16 shows a clutch of this type used for a starter drive. When the engine starts, the coupling member that is attached to the engine shaft overtakes the coupling member connected to the starter motor. Because of the shape of the coupling teeth, the coupling member is pushed out of engagement.

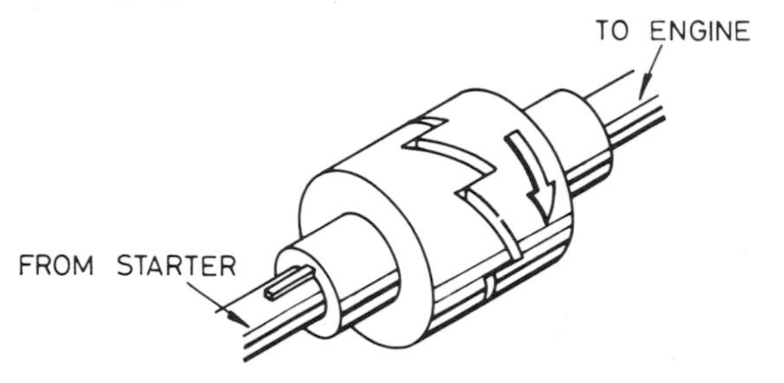

Fig. 12.16 Claw coupling

(d) Shear link

A shear link is included when the connection must be broken, to prevent damage to the mechanism, in the event of it becoming overloaded. A typical shear link consists of a flanged coupling in which the connecting bolts are waisted so that they will shear on the joint plane between the flanges when the overloading conditions are reached.

12.3 Transmission of rotational motion using shafts that are not coaxial

Shafts whose axes are not coaxial will form part of a transmission system if the driving motor must be separated from the output, if it is inconvenient to position the motor and the output in line, or if the input and output shafts must rotate at different speeds. The system used to connect the shafts depends upon their rotational speeds, the distance between their axes and their relative positions, the magnitude of the power to be transmitted, the space that is available for the device and the extent to which maintenance can be done. The choice is further limited if there must be a fixed relationship between the angular positions of the two shafts (for example, the driving system used for timing in a petrol engine).

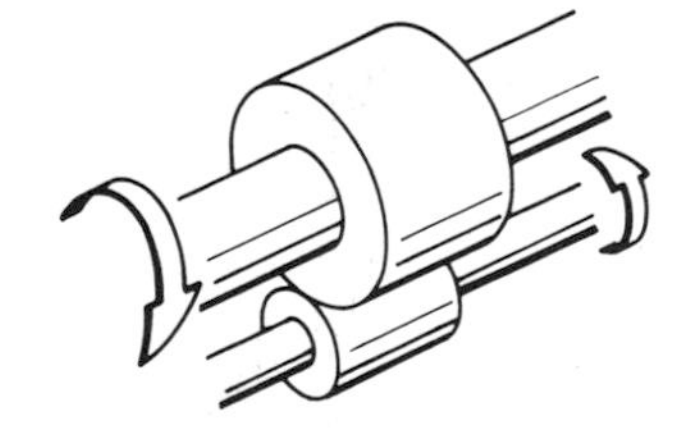

Fig. 12.17 Friction-disc drive

(a) Friction-disc drive

Fig. 12.17 shows a simple friction drive. The ratio of the speeds of the two shafts depends upon the ratio of the diameters of the friction discs. In the arrangement shown, the shafts will rotate in opposite directions; but an idler disc between them would reverse the direction of rotation of the driven disc without altering the ratio of the speeds of the input and output shafts. Fig. 12.18 shows a friction-disc transmission used to connect shafts whose axes are at right angles and intersect.

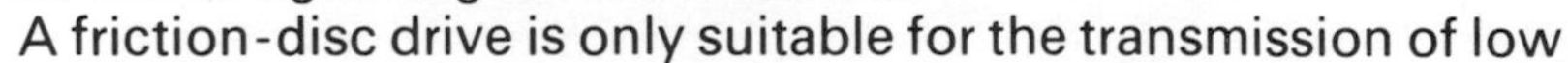

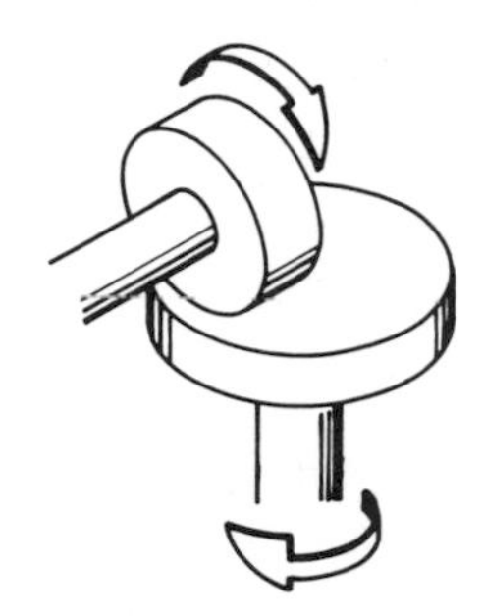

Fig. 12.18 Friction-disc drive

A friction-disc drive is only suitable for the transmission of low power. The relationship between the speeds of the shafts tends to vary because of slip and the angular relationship between the shafts is not fixed. The system is not used to connect shafts whose axes are far apart because the discs would need to be of large diameters to 'make up' the gap and the drive would, as a result, require a fairly large amount of space.

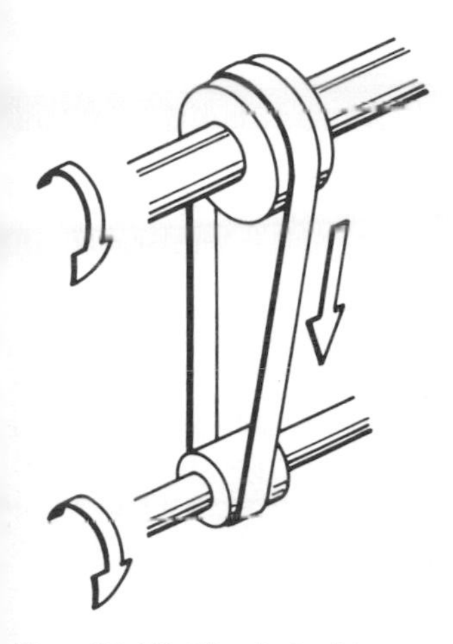

Fig. 12.19 Flat-belt drive

(b) Belt drive

A belt drive can be used to connect shafts that are far apart without taking up too much space; but the shafts must be far enough apart to allow space for the two pulleys over which the belt passes. The belt can be flat (as shown in fig. 12.19), be of *vee* section, or be of circular section. When the belt is in the *open position* (see fig. 12.19) the pulleys will rotate in the same direction, but by crossing the belt they can be made to rotate in opposite directions. The ratio of the shaft speeds depends upon the ratio of the diameter of the two pulleys, but the speed ratio tends to vary slightly because of belt slip. (Slip is reduced by using a vee belt—but a flat belt is often used in a drive so that slip can prevent damage to the motor in the event of overloading.) Belts and pulleys are robust and belts can run at speeds of up to about 26 m/s, but the system needs clean conditions and some maintenance. Plain belts do not give a fixed relationship between the shafts, but this can be obtained by using a toothed-belt system as shown in fig. 12.20.

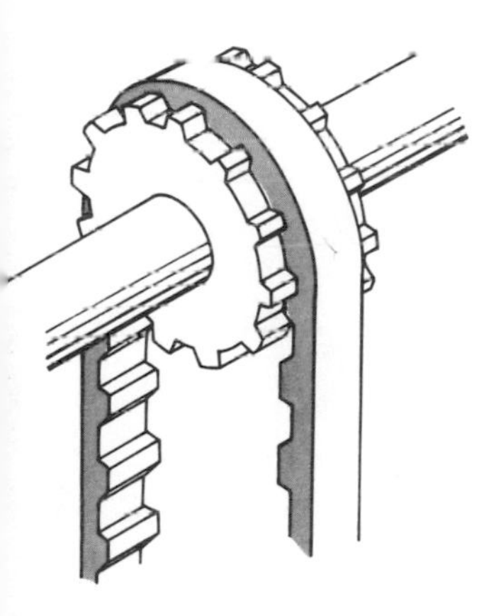

Fig. 12.20 Toothed-belt drive

(c) Chain drive

A chain drive is similar to a belt drive but it enables a higher power to be transmitted without slip. Also it maintains a fixed relationship between the angular positions of the two shafts. The ratio of the shaft speeds depends upon the diameter of the two sprockets. A chain-drive system requires lubrication and some maintenance.

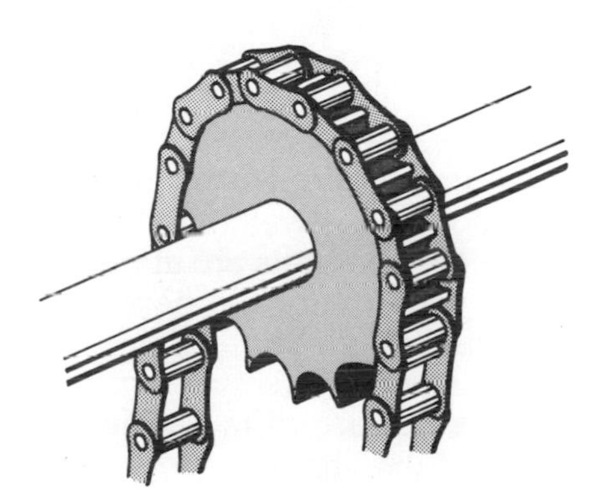

Fig. 12.21 Chain drive

(d) Toothed-gear drive

High speeds and high power can be transmitted using toothed gears instead of friction discs (compare fig. 12.22 with fig. 12.17). In this system there is no slip but there is some backlash unless gears of a special design are used. Unlike a friction-disc drive, a toothed-gear drive provides a fixed angular relationship between the shafts; but, like the friction-disc drive, the distance between them is limited unless a gear train is used. The drive must be lubricated and some maintenance is required. A gear drive using gears with straight teeth is not as quiet or as smooth as a belt drive, but this can be improved, at extra cost, by making the gear teeth helical (compare figs 12.23 and 12.24). Helical gears produce axial thrust (the magnitude of which increases with the helix angle), but this can be nullified by using double-helical gears.

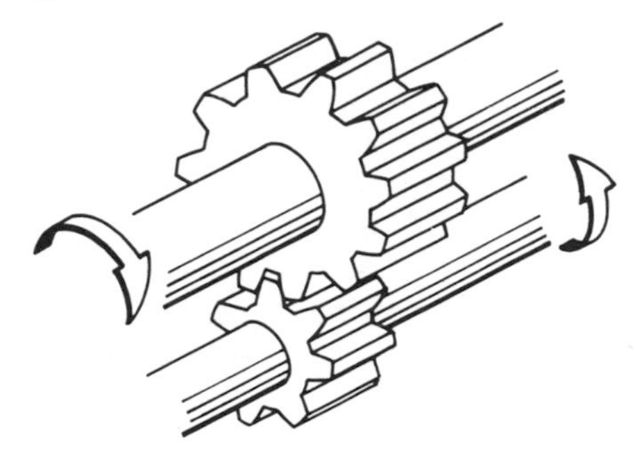

Fig. 12.22 Straight spur-gear drive

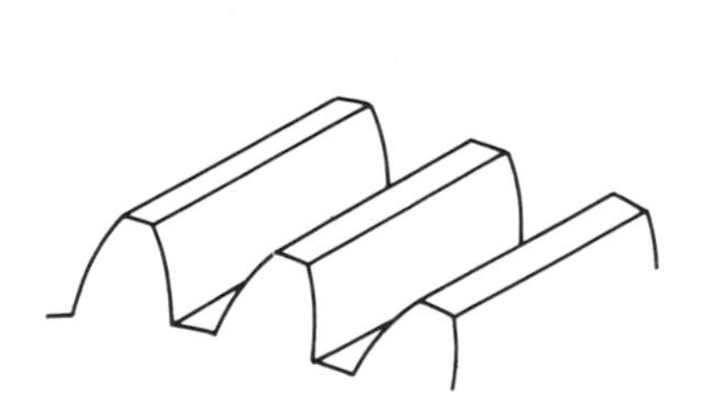

Fig. 12.23 Straight gear teeth

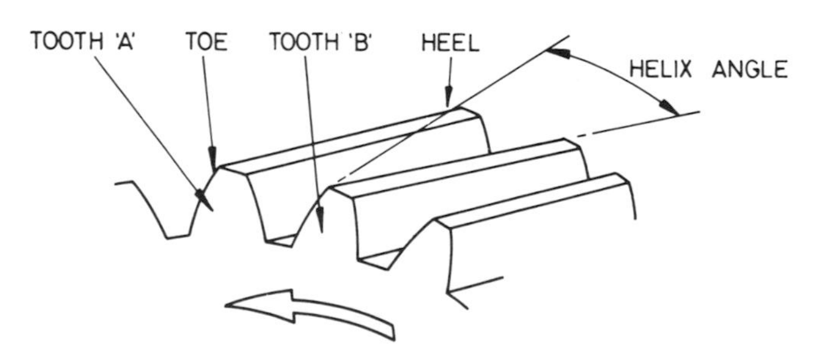

Fig. 12.24 Helical gear teeth

Shafts whose axes are at right angles can be connected by a bevel-gear drive. The teeth can be straight (as illustrated in fig. 12.25), or spiral (equivalent to helical gears) to cope with higher speeds. When the shafts are at right angles but their axes do not intersect, a variation of the bevel-gear system, called a hypoid-gear system, is used. Shafts whose axes cross can be connected by skew or spiral gears (see fig. 12.26). These gears resemble helical gears but are of larger helix angle to accommodate the inclination of the shafts.

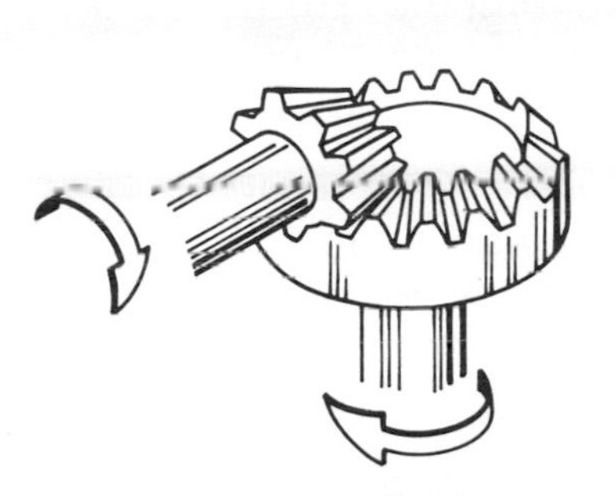

Fig. 12.25 Bevel-gear drive

Fig. 12.26 Spiral-gear drive

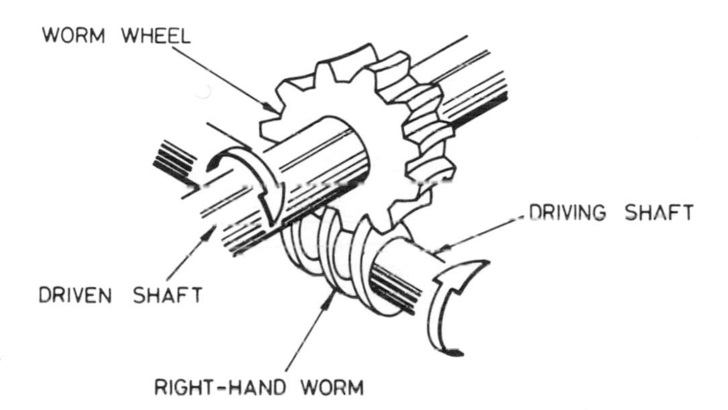

Fig. 12.27 Worm and worm-wheel drive

A worm and worm-wheel drive, as shown in fig. 12.27, is used to connect shafts that cross at right angles. This type of gear drive is effective when a high reduction is required in a relatively small space.

(e) Crank and coupling-rod drive

Fig. 12.28 shows this simple system, which gives a fixed relationship between the two or more shafts it connects but usually takes up a large amount of space. This system was used to couple the driving wheels of steam locomotives.

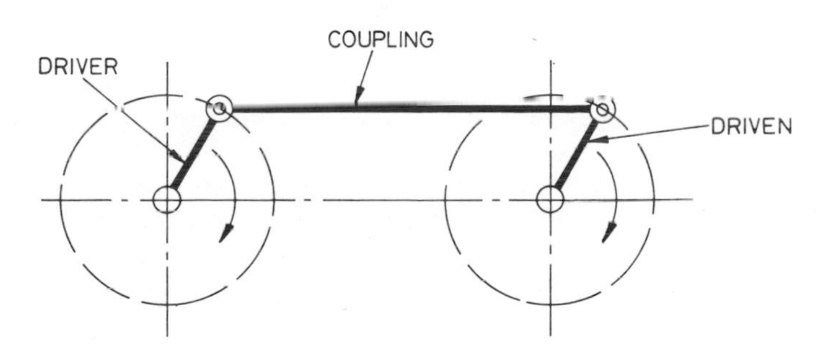

Fig. 12.28 Crank and coupling-rod drive

(f) Mechanisms to produce an interrupted circular motion

Sometimes a shaft that is driven by a shaft that is continuously rotating must have an interrupted circular motion. A typical application of such a drive can be seen in a movie projector, in which each frame in turn must be held steady for a very short period of time.

Figs 12.29 and 12.30 illustrate a Geneva mechanism designed to cause the driven member to rotate in 90° increments, two of which take place during one revolution of the driver, and to be located during the dwell period. In this example, the driven member has four slots (to produce the 90° increments) and the driver has two driving pins (to cause the driven member to move twice during each revolution of the driving shaft). The raised plates on the driver engage in the scallops in the driven member to locate it during each dwell.

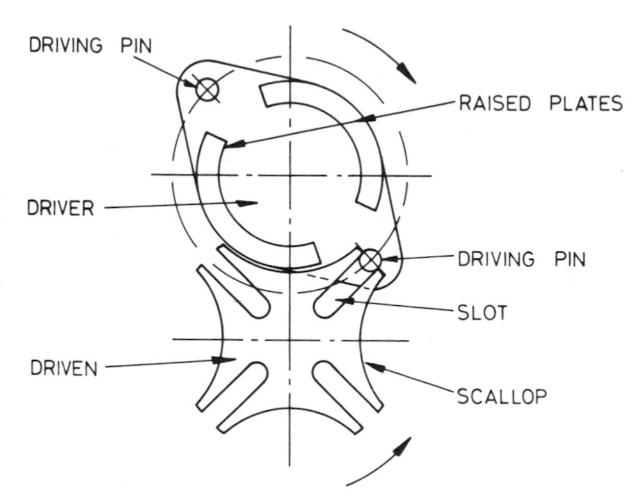

Fig. 12.29 Geneva mechanism

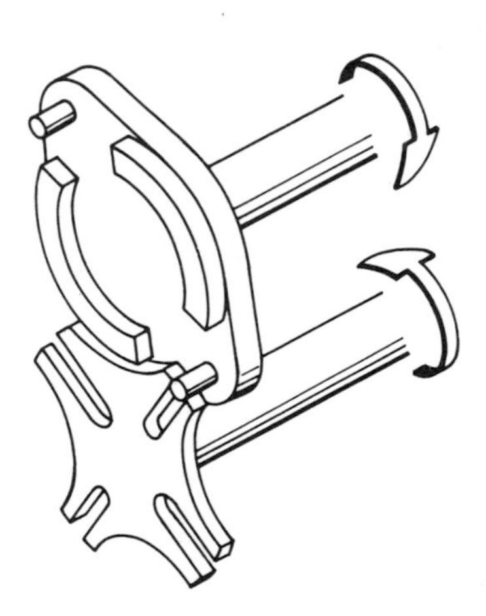

Fig. 12.30 Geneva mechanism

An interrupted motion can also be obtained using a gear mechanism as shown in fig. 12.31. The two gears that form the drive have teeth only on part of their circumference, so that the driven gear will be stationary when the teeth are not in engagement but will rotate when the driver teeth engage its teeth and will continue to rotate until the teeth disengage. The illustration shows the driven gear in a stationary position after the driver teeth have disengaged teeth A on the driven gear (the driven gear is located during the stationary period by the circular portion of the driver gear). When the driver teeth engage teeth B on the driven gear, the motion of the latter will be resumed. The arrangement of the teeth on the driven gear controls the angular increments and the gear ratio and arrangement of the teeth on the driver gear controls the frequency of the movements.

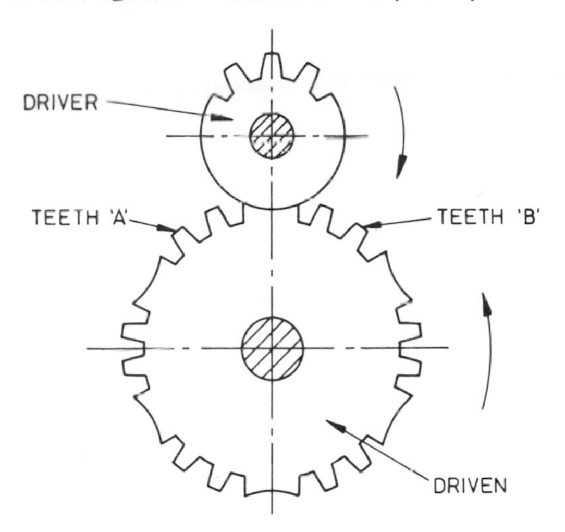

Fig. 12.31 Interrupted gear drive

12.4 Conversion of rotational motion into reciprocating motion perpendicular to the axis of rotation

The mechanism that is selected to produce the reciprocating motion depends upon the characteristics of the motion (for example, uniform velocity or varying velocity, periods of dwell etc.), the required degree of control over the driven member, the magnitude of the reciprocating movement and the space that is available for the mechanism. The principal mechanisms and their characteristics are described in this section.

(a) Mechanisms based on the slider-crank chain

Fig. 12.32 shows the elements of the slider-crank chain. The slider is attached at point A to the link AC so that it can turn, and slides in the slotted link OB. The third link OC is attached to link AC at point C and to link OB at point O so that turning can occur at each of these points. A mechanism is produced by fixing one of the links of this kinematic chain.

Fig. 12.33 illustrates the slider-crank mechanism that is produced by fixing link OB by making it part of the machine body, so that as the link OC (then called the *crank*) rotates about O, the connecting rod (link AC) causes the slider to reciprocate along the line OB; alternatively, movement of the slider will cause link OC to rotate about point O. A uniform velocity of the link OC will not produce a uniform velocity of the slider (and a uniform velocity of the slider will not produce uniform velocity of link OC)

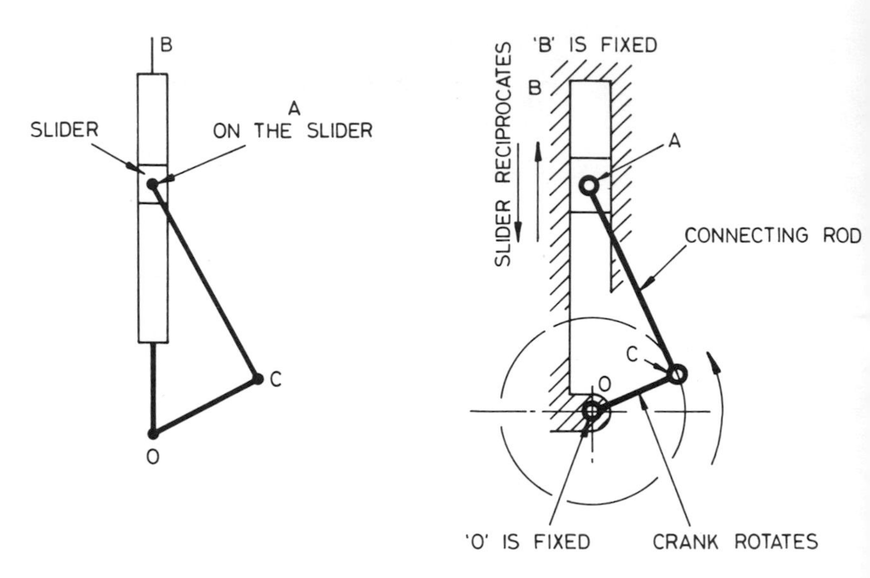

Fig. 12.32 Slider-crank chain

Fig. 12.33 Slider-crank mechanism

because of the geometry of the mechanism. A dwell or a reversal of the direction of movement cannot be obtained until the end of the stroke is reached (except by changing the driver movement) and the crank requires a lot of space compared with the length of the stroke it produces, but the slider is positively controlled throughout the whole cycle. The piston–connecting-rod–crankshaft mechanism in an internal combustion engine is a familiar application of this mechanism in which the piston (the slider) drives the crank during the firing stroke and the crank drives the piston during the remaining three strokes. As the stroke is short the crank swing is small and the cylinder block and cylinder can be designed to accommodate the connecting-rod. When the connecting-rod cannot enter the cylinder, the slider can be extended to produce the cross head, piston-rod and piston arrangement shown in fig. 12.34.

Another mechanism based on the slider-crank chain is used in machine tools of the reciprocating type to minimise the non-

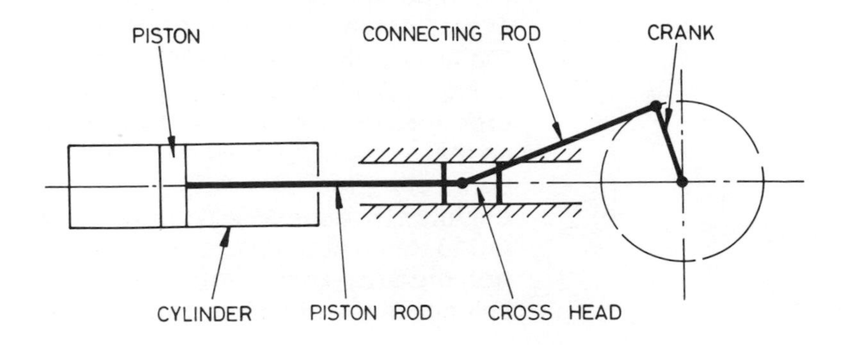

Fig. 12.34 Slider-crank mechanism with cross head

cutting time by causing the ram to which the cutting tool is attached to move at a greater average speed during its return (non-cutting) stroke than it does during the cutting stroke. The Whitworth quick-return mechanism, shown in simplified form in fig 12.35, is suitable for slotting machines. It will be seen from fig.

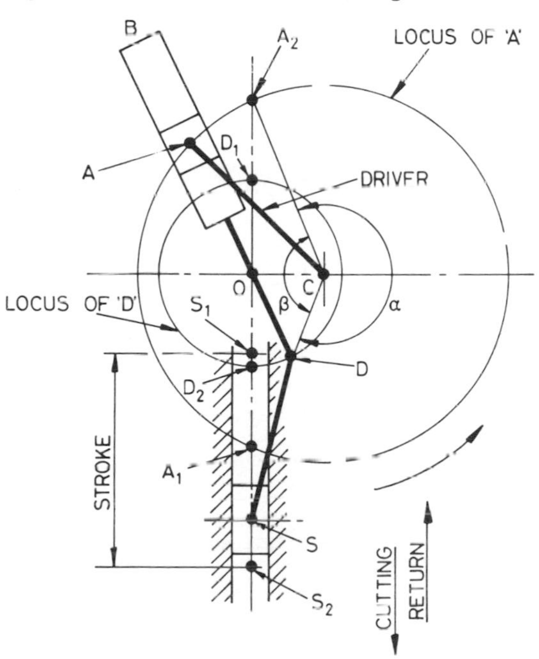

Fig. 12.35 Whitworth quick-return mechanism

12.36(a), which shows part of the mechanism, that link OC is fixed and that the driver link, AC, drives the slotted link, OB. The slider is free to rotate about the pin, A, and to slide in the slot so that it can assume its required position and attitude. The side view, fig 12.36(b), shows that this linkage is a coupling between the driver shaft and the driven shaft. Point A_1 and point A_2 in fig. 12.36(a) represent, respectively, the 'bottom dead centre' and the 'top dead centre' positions of pin A. It will be seen that the driver link AC rotates through a larger angle (α)to cause A to move from A_1 to A_2 than it does to cause it to return from A_2 to A_1 (angle β). When link AC rotates at a constant speed, the time taken for A to move from A_2 to A_1 is less than that for it to move from A_1 to A_2; this is the basis of the quick-return action. Fig. 12.36(c) and fig. 12.36(d) shows that the rotational motion of link OB can produce a reciprocating movement by attaching a crank to the end of the driven shaft and linking it at D to a connecting-rod which, in turn, is attached at S to the machine ram which moves in a slide. The crank, OD, is, in effect, an extension of the link OB. S_1 and S_2 represent the extreme positions of point S on the machine ram, and it will be seen from fig. 12.36(c) that S will be in its bottom position, S_2, when A is at position A_2, and be in its top position, S_1, when A is at position

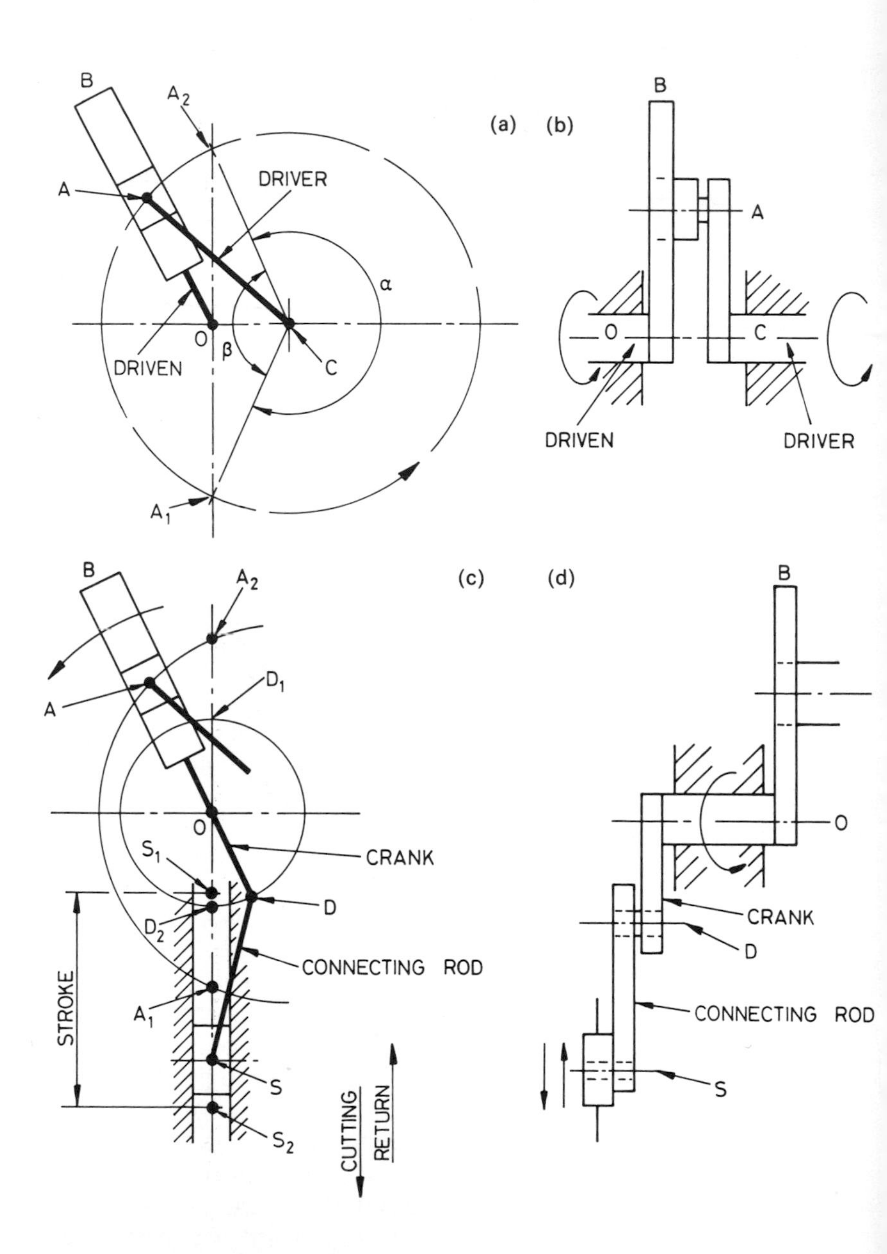

Fig. 12.36 Parts of the Whitworth quick-return mechanism

A_1. It has already been explained that A moves from position A_2 to position A_1 (when S is moving upwards on the return stroke) quicker than it moves from position A_1 to position A_2 (when S is moving downwards on the cutting stroke) and so a quick-return action is obtained.

A variation of this mechanism is used in shaping machines to obtain a longer stroke with an arrangement that is compact and fits into the body of the machine. In this variation, shown in fig. 12.37 and fig. 12.38, OC is again fixed, but its length is such that point O lies outside the locus of pin A on the link AC and, as a result, link OB is given an oscillating motion. Points B_1 and B_2 represent the extreme positions of the link-pin B and points A_1 and A_2 are the positions of pin A when B is at its extreme positions. It will be seen that AC rotates through a larger angle (α) when A moves from A_1 to A_2 than it does when it moves from A_2 to A_1 (angle β), and so when AC rotates at a constant speed the time taken for B to move from B_2 to B_1 will be less than that

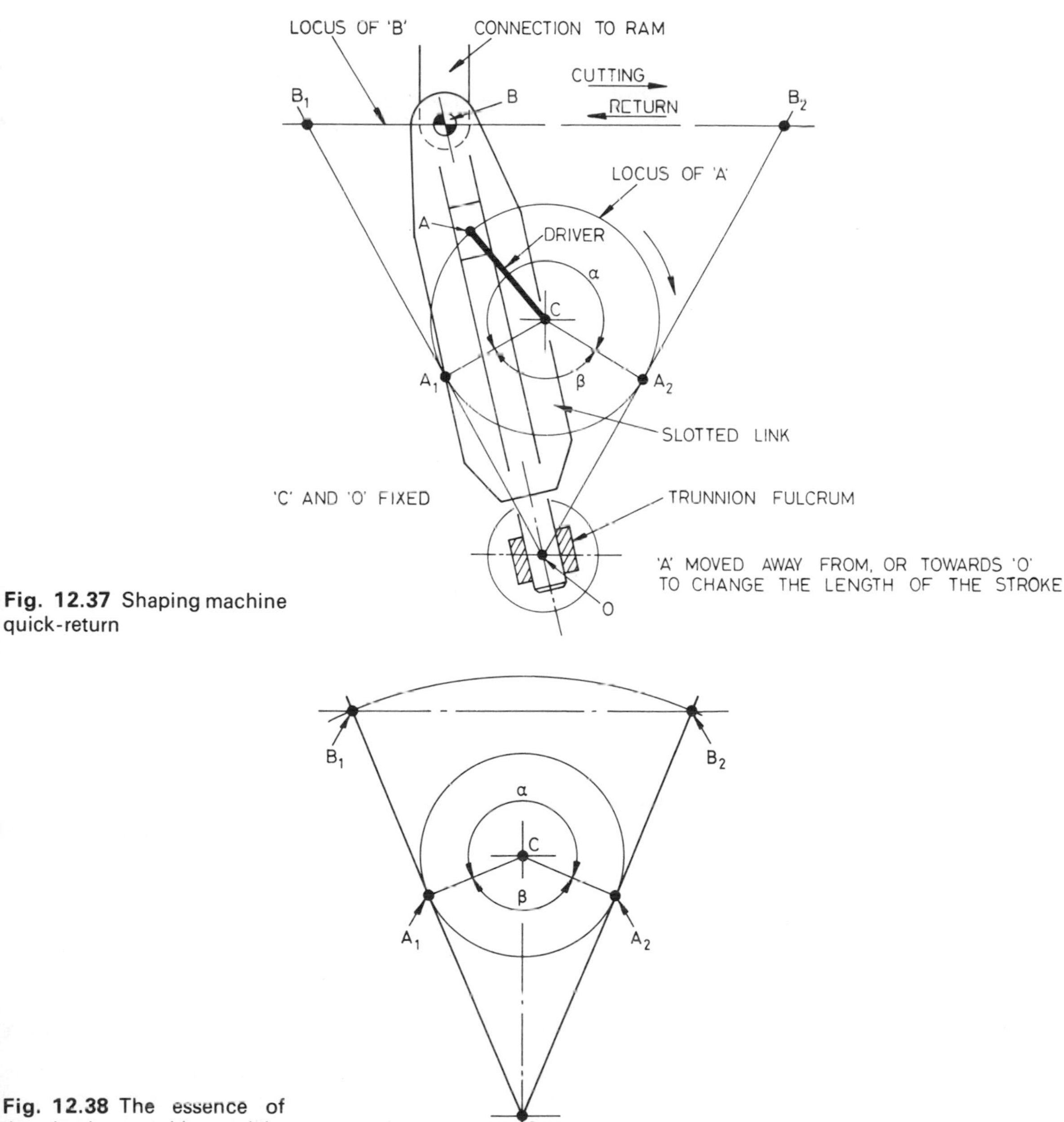

Fig. 12.37 Shaping machine quick-return

Fig. 12.38 The essence of the shaping machine quick-return

for it to move from B_1 to B_2. This is the basis of the quick-return action. It will be seen from fig. 12.38 that if link OB is fixed at O, the locus of the link-pin B will be an arc instead of the straight line that is required to suit the ram movement. Because of this the end of the link OB is located in a trunnion, as shown in fig. 12.37, so that the link-pin B can assume its required position. The stroke of the ram is changed by moving A towards or away from C.

Unfortunately the geometry of both these mechanisms is such that the cutting speed changes throughout the stroke. The rate of variation is greater at the extremes of the stroke that at the middle section but often it is possible to set the machine and the workpiece so that cutting takes place along the portion of the stroke where the cutting speed is reasonably uniform.

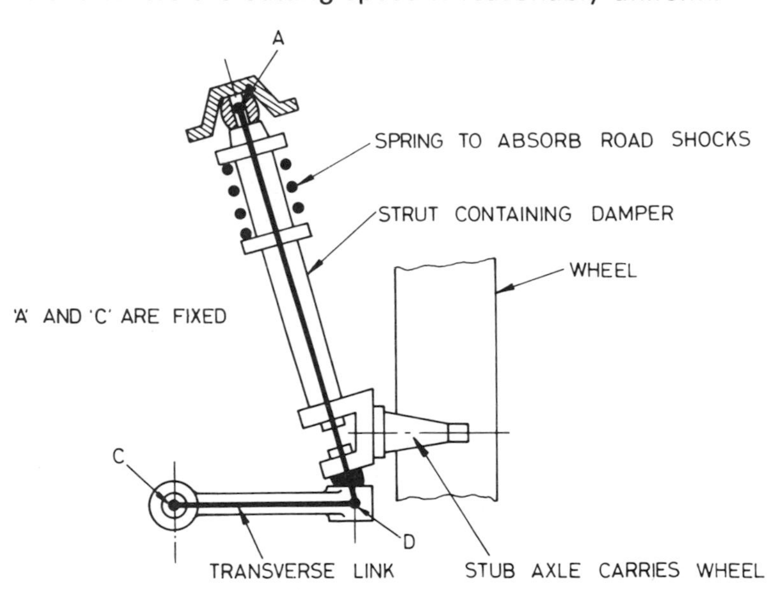

Fig. 12.39 MacPherson strut assembly

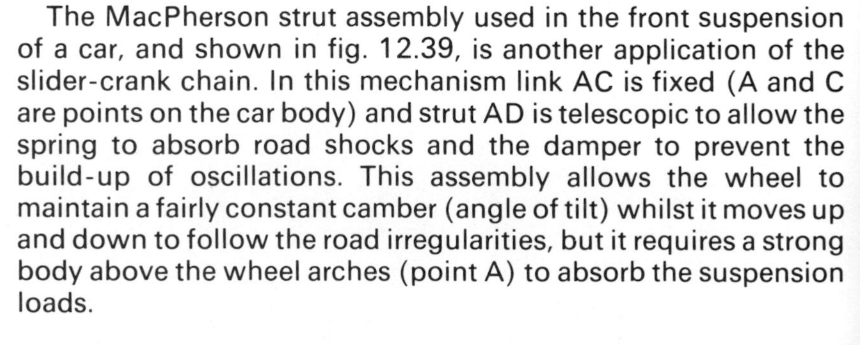

The MacPherson strut assembly used in the front suspension of a car, and shown in fig. 12.39, is another application of the slider-crank chain. In this mechanism link AC is fixed (A and C are points on the car body) and strut AD is telescopic to allow the spring to absorb road shocks and the damper to prevent the build-up of oscillations. This assembly allows the wheel to maintain a fairly constant camber (angle of tilt) whilst it moves up and down to follow the road irregularities, but it requires a strong body above the wheel arches (point A) to absorb the suspension loads.

(b) Mechanisms based on the double slider-crank chain

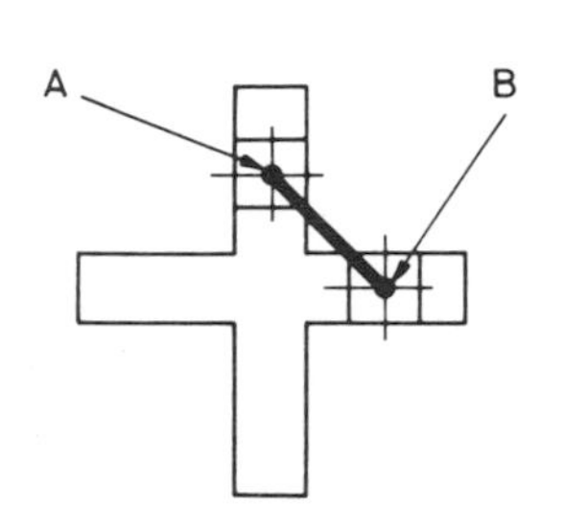

Fig. 12.40 Double slider-crank chain

This chain, illustrated in fig. 12.40, consists of two sliders, linked together at A and B, that move in two slots. If one slider (say the one attached to link AB at point A) is fixed, the slotted frame will reciprocate when link AB is rotated about B. The Scotch Yoke

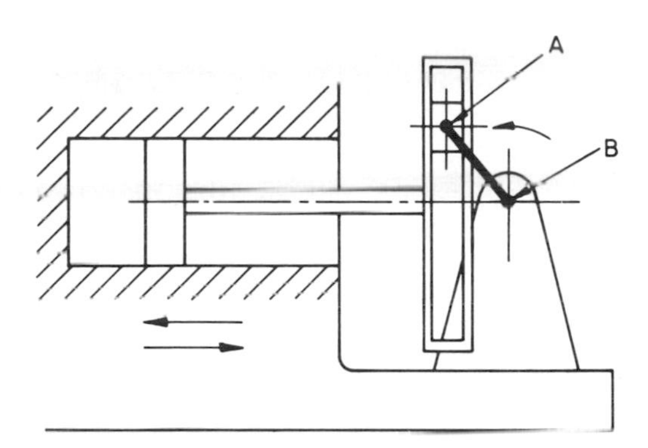

Fig. 12.41 Scotch yoke

(fig. 12.41) is a practical version of this mechanism which eliminates the connecting-rod associated with the slider-crank mechanism in fig. 12.33.

If link AB is fixed, rotation of the slider at point B about that point will cause the slotted frame to rotate through the same angle. The frame will, in turn, cause the other slider to rotate about A through the same angle. This is the basis of the Oldham coupling described in section 12.2(a) (fig. 12.12).

(c) Cam and follower

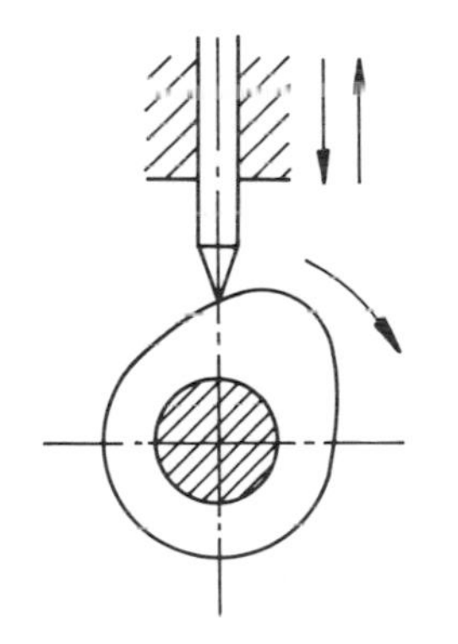

Fig. 12.42 Cam and knife-edge follower

Figs 12.42 to 12.44 show three cam-and-follower mechanisms. In all the arrangements the follower movement can, by use of a suitable cam profile, be made to vary, both in speed and direction, throughout the cycle, but the actual movement that can be obtained is limited by the type of follower that is used. The knife-edge follower (fig. 12.42) can follow a complicated cam profile but is prone to wear. Wear can be reduced by using a flat-faced follower as shown in fig. 12.43 because contact between it and the cam edge occurs at differing places on the follower face, so that wear is not localised. The follower movement that is obtained by this system is limited because the cam profile must be convex at all places to enable the follower to maintain correct contact with it. The roller follower (fig. 12.44) is not prone to wear, but the cam profile is limited because the diameter of the roller must be large enough to enable a reasonably large roller-axle to be used, and the concave regions of the cam profile must be of a radius that is no smaller than that of the roller, so that full contact can be maintained.

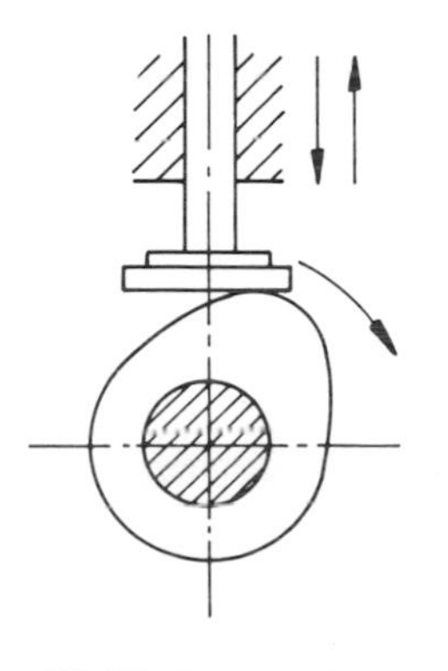

Fig. 12.43 Cam and flat-faced follower

The arrangements described so far may not provide positive control of the follower even when its movement is in a vertical direction (and contact between it and the cam profile is assisted by gravity) because the lift motion imparted to the follower may cause it to accelerate away from the cam profile. In addition to producing an incorrect movement this will cause wear of the cam profile and of the follower because of the hammering action each time the cam profile catches up with the follower. Similarly the

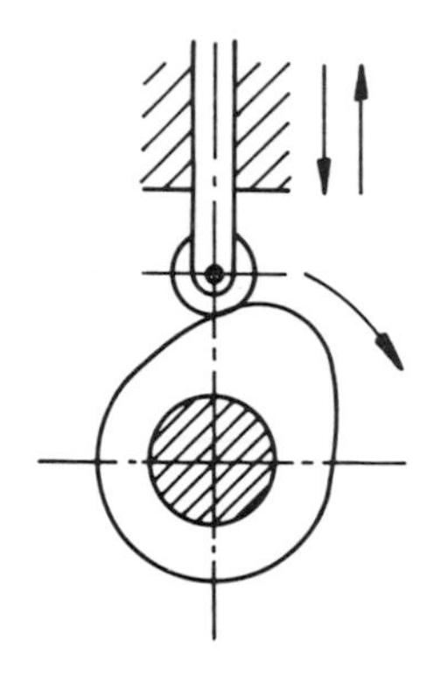

Fig. 12.44 Cam and roller follower

cam profile may be such that the follower cannot maintain a continual contact with it during the fall, and hammering occurs when contact is regained. When the output includes a spring (for example, the valve-operating system of an internal combustion engine) or one can be introduced to press the follower against the cam, the contact may be maintained; but very often the cam profile may need to be modified to smooth out the movement because of the tendency for bouncing to take place. Alternatively, a cam system that uses a cam track and a double-roller follower, as shown in fig. 12.45, can be used instead of an edge cam.

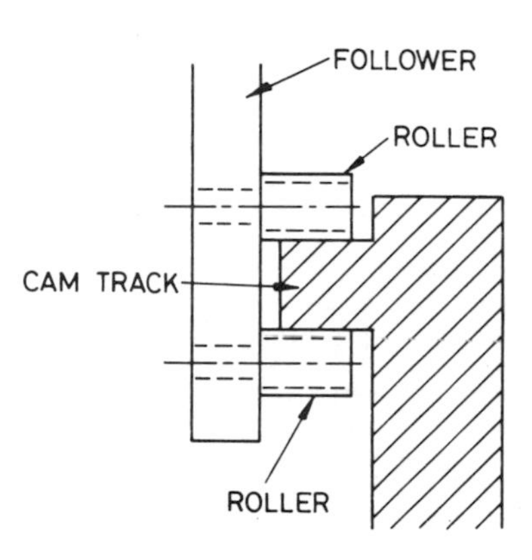

Fig. 12.45 Cam track and double-roller follower system

(d) Eccentric and follower

When a follower is to be given a simple movement (for example, to open and close a valve, to operate a pump, or to control the stroke of a metal-working press) an eccentric may be used instead of the more expensive cam (see fig. 12.46). Several eccentrics may be mounted on a driving shaft, and the movements produced by them synchronised by rotating them in turn about the shaft axis before locking them in position. The characteristics of this mechanism are similar to those of the cam and follower.

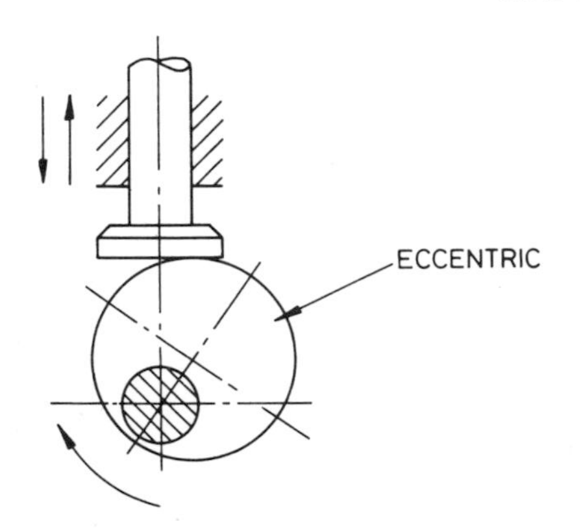

Fig. 12.46 Eccentric and follower

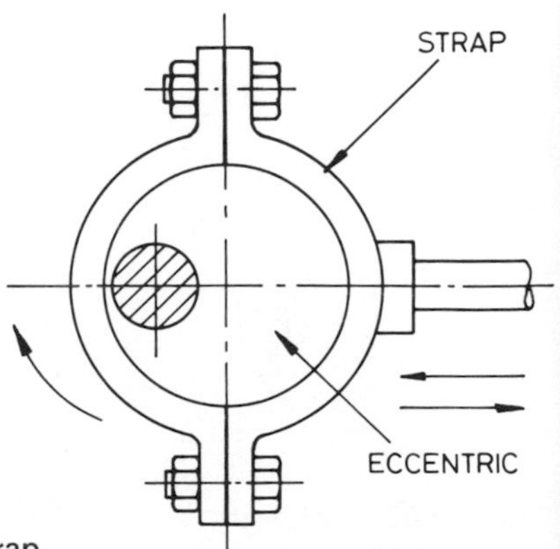

Fig. 12.47 Eccentric and strap

A positive movement of the follower can be obtained by enclosing the eccentric with a strap as illustrated by fig. 12.47.

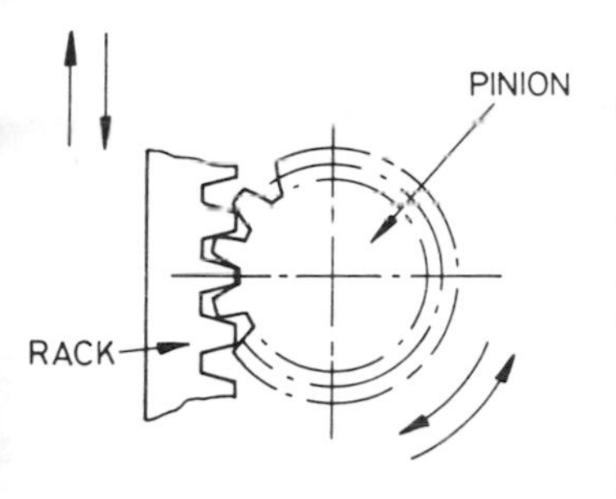

Fig. 12.48 Rack-and-pinion drive

(e) Rack-and-pinion drive

A positive, uniform movement is produced by a rack-and pinion mechanism (fig. 12.48). The ratio of the pinion and rack speeds depends upon the diameter of the pinion. A reciprocating motion of the rack cannot be obtained without reversing the direction of rotation of the pinion, but a modified version of the system with two racks, one on each side of the pinion, which engages them in turn, will produce a reciprocating action.

12.5 Conversion of rotational motion into reciprocating motion parallel to the axis of rotation

As in the previous section, the mechanism that is selected depends upon the characteristics of the motion, the required degree of control over the driven member, the magnitude of the movement and the space that is available. The principal mechanisms and their characteristics are described in this section.

(a) Cylindrical cam and follower

The system shown in fig. 12.49 gives a reciprocating motion with positive control to the follower (it is similar in this respect to the system in fig. 12.45). The movement is limited by the follower shape and its ability to follow the cam track. The Bell cam, shown in fig. 12.50, is similar to the edge cams (see figs 12.42 to 12.44) and almost always requires the application of a force to cause the follower to maintain contact with the cam track because the camshaft axis is usually horizontal.

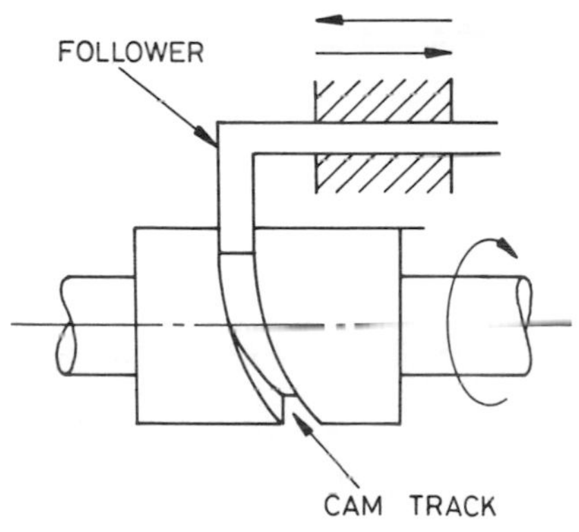

Fig. 12.49 Cylindrical cam and follower

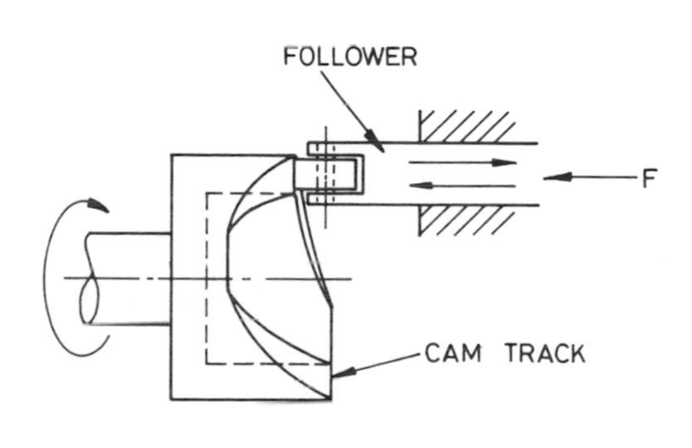

Fig. 12.50 Bell cam and follower

(b) Swash plate and follower

This system (fig. 12.51) is similar in principle to systems that use an eccentric. It is useful when, for example, a pump is to be operated by a simple reciprocating movement.

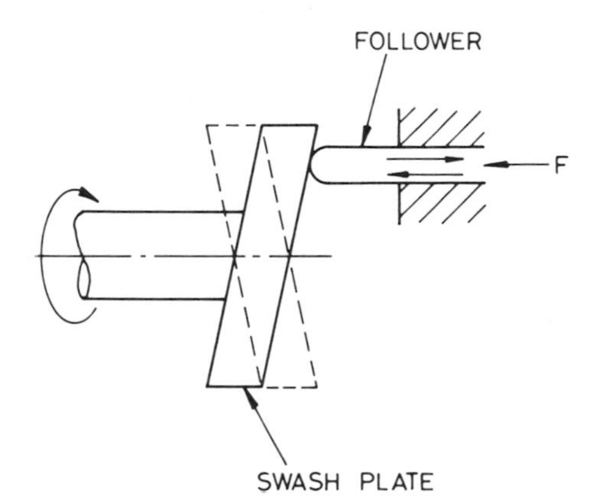

Fig. 12.51 Swash plate and follower

(c) Worm and rack

A positive linear motion is produced by this mechanism (fig. 12.52). It is similar to the rack-and-pinion system in that it requires the reversal of the worm to move the rack in the opposite direction. A high rack-to-shaft speed ratio can be obtained by using a worm with a large lead. This system is usually unsuitable when a massive component, such as a machine table, is to be moved because the force is not exerted along the axis of the worm shaft.

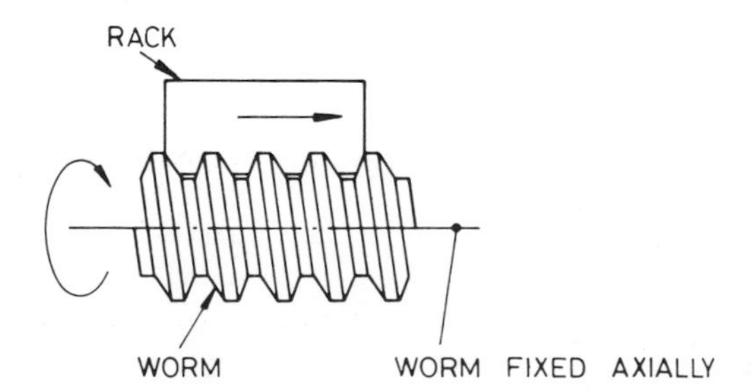

Fig. 12.52 Worm and rack

(d) Leadscrew and nut

This mechanism is closely related to the worm and rack, but the force is exerted in a better way because the leadscrew is encircled by the nut. The leadscrew may be constrained axially so that the nut, which cannot rotate, will move along it (this arrangement, which is shown in fig. 12.53, is used in a lathe). Alternatively the nut may be completely constrained so that the leadscrew moves axially whilst it rotated (this arrangement, shown in fig. 12.54 is used in a milling machine).

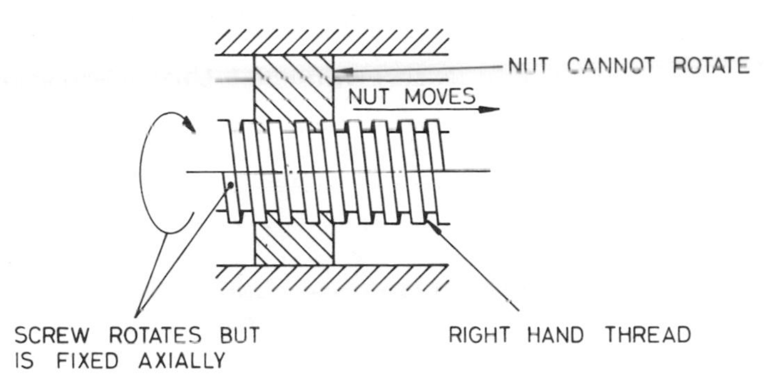

Fig. 12.53 Leadscrew and nut

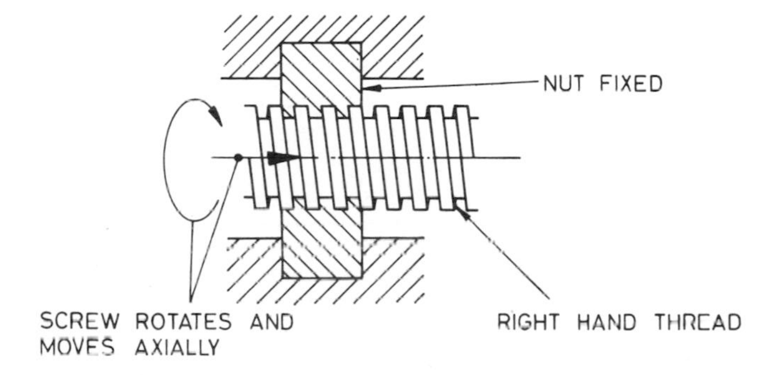

Fig. 12.54 Leadscrew and nut

12.6 Pawl-and-ratchet mechanisms

These mechanisms consist of two principal components: the pawl and the ratchet wheel. The system is designed so that relative movement in one direction causes the ratchet to engage the ratchet wheel teeth, but to ride over them when the direction of movement is reversed. Fig. 12.55 shows such a system used to produce incremental movement of the ratchet wheel. The pawl is mounted on an arm which oscillates as indicated and produces an

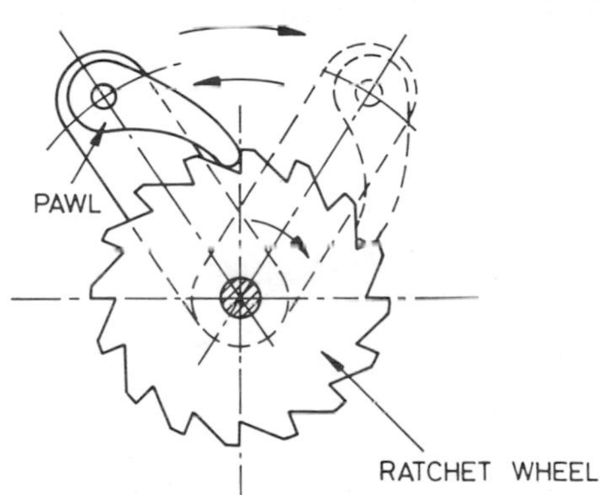

Fig. 12.55 Pawl-and-ratchet mechanism

incremental rotational movement of the ratchet wheel when moving clockwise by engaging its teeth, but rides over them when turning anti-clockwise. (This system is used to produce a side feed in machine tools such as a shaping machine, and in ratchet screwdrivers.) The system shown in fig. 12.56 allows the ratchet wheel to be rotated by applying a force against an opposing force which may either be prevented from turning the ratchet wheel when the applied force is withdrawn (a winch uses this system) or allowed to rotate a wheel upon which both the ratchet wheel and the pawl are mounted (a clock spring winding system is an example of this system). A fixed pawl (as in fig. 12.56) can be included in the mechanism shown in fig. 12.55 if the ratchet wheel is likely to rotate in an anti-clockwise direction when it is not being moved by the oscillating pawl.

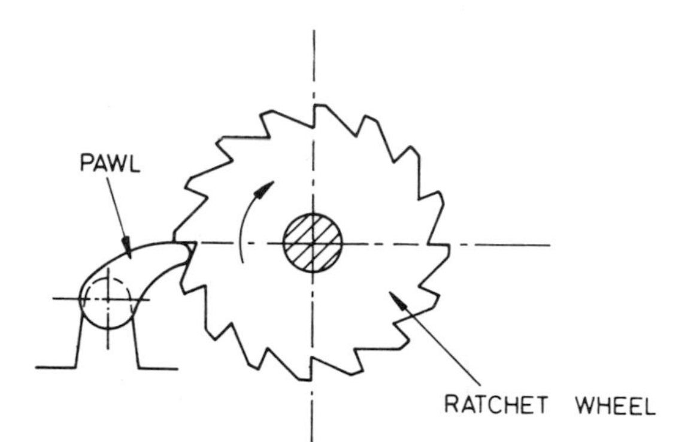

Fig. 12.56 Pawl-and-ratchet mechanism

12.7 Toggle mechanisms

This mechanism is illustrated in fig. 12.57. Link AB is attached to link OA at point A, and to link BP at point B so that turning can occur at these points. Link OA can turn about the fixed point O and link BP moves along the line OP. When link OA is turned so that point A moves along arc XY, point B moves away from O. It will be seen that equal movements of point A will not produce equal movements of point B, and that the movements of B that

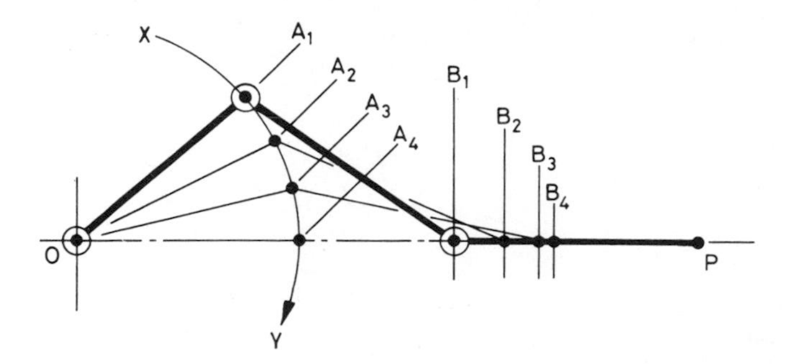

Fig. 12.57 Toggle mechanism

are produced become smaller as A moves from A_1 to A_4. This means that starting from A_1, when the velocity ratio is high, the situtation changes so that as A_4 is approached, the mechanical advantage becomes high and the velocity ratio becomes small. A typical application of a toggle mechanism will be seen in a press where the ram approaches the workpiccc at high speed but exerts a high force on it at the point of engagement. Fig. 12.58 shows a toggle clamp. The plunger is adjusted (by a system that is omitted from this diagram) so that the clamping action is completed when joint A is over dead centre. This produces a self-locking action because to release the clamp link OA must be turned in an anti-clockwise direction—but can only be turned by increasing the clamping force on the workpiece.

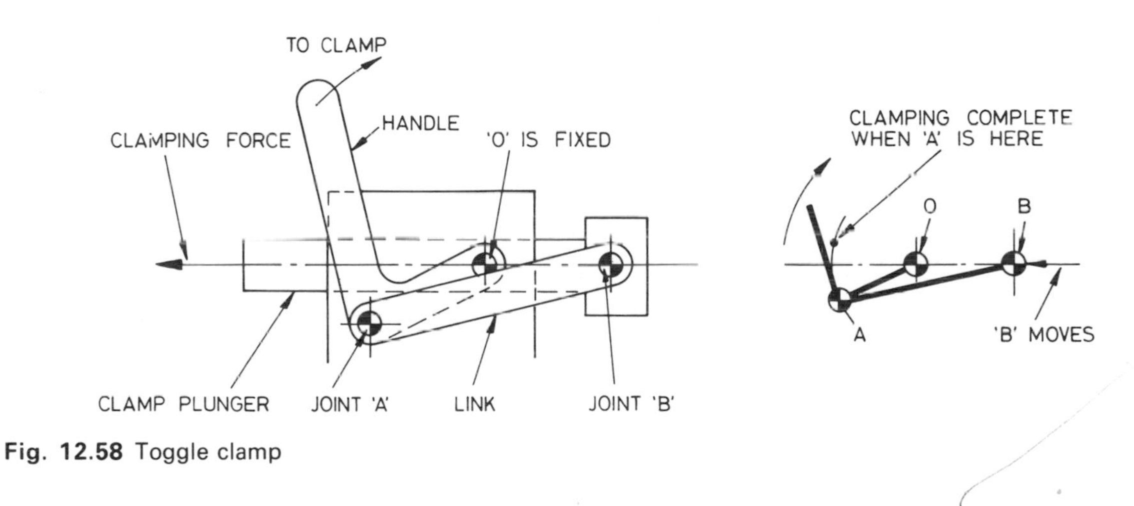

Fig. 12.58 Toggle clamp

12.8 Safety

It is the duty of the designer to consider the safety of the user and of other people in the vicinity of mechanisms. The extent to which safety becomes a problem depends upon the circumstances in which a mechanism is used. For example, when gears must be enclosed in a gearbox casing to function as required, they are unlikely ever to be a source of danger; but if they do not need to be enclosed, they are a potential danger.

Rotating parts should, if not enclosed in a casing, be surrounded with guards to prevent people or their clothing from being caught up in the mechanism, or from being injured as a result of parts of it working loose or breaking. When the user is required to use a mechanism that cannot be guarded all the time it is in use, the system must be such that the mechanism will not operate when the user is in danger. For example, when the user must place each workpiece in the die of a press tool set, it must be impossible for the press to close until the user's hands are in a safe position, and the guard is in place.

Controls should be located so that the user is not required to lean across a mechanism, and an emergency control or emergency controls should be located to enable the mechanism to be stopped immediately in the event of an accident. Care must be taken that an opposing force cannot act against the user so as to be dangerous; the use of a ratchet in a winch is a typical example of a safety system of this type.

In general, the application of ergonomics in conjunction with the design of mechanisms that are themselves safe, will ensure the overall safety of a system.

13 Dimensional and Geometrical Tolerancing

In order that a component will fit with other components in an assembly, perform as required, and in order that it can be replaced, when necessary, by another component made from the same drawing (a system known as interchangeability), it is necessary for it to be of an acceptable size and shape and have the required surface texture. As it is impossible to achieve perfection, the permissible errors in its manufacture must be determined and stipulated on the drawing or in other instructions. The permitted error is called the tolerance, and the extreme conditions of size, shape and surface texture are called the limits. The magnitude of the tolerance depends upon the method of manufacture, the accuracy of the manufacturing equipment, the skill of the manufacturer, the size of the component (for a given quality of accuracy, the tolerance usually increases with size) and the allowable cost of the component.

The determination and stipulation of the tolerance may be considered in three sections. These are:

- size (limits and fits);
- shape (geometrical tolerancing);
- surface texture.

Size and shape are considered here and the specification of surface texture is the subject of chapter 14.

13.1 Limits and fits

It is necessary to consider the size of the individual components and, when they are required to fit together, the relationship between the size of the parts.

(a) Unilateral limits and bilateral limits

Unilateral limits are usually used when the distance between two faces, or the diameter of a hole or shaft is specified. For example, when external-grinding a shaft, the machinist would prefer to aim at the largest size permitted, so that, in the event of him reaching a diameter that is just a little larger then the maximum size permitted, he can take another cut, knowing that he can use up the whole of the tolerance before the job is rejected. A draughtsman might dimension a nominal 75 mm diameter shaft as: ϕ 75·012–0·012, or, alternatively, as $\phi\ {}^{75.012}_{75.000}$. The advantage of the first method is that the size to aim at is clearly indicated, but the advantage of the second method is that there is no chance of the operator making an arithmetical error when working out the unspecified limit. Similarly, a nominal 75 mm hole might be dimensioned as ϕ 75 + 0·012, or alternatively as $\phi\ {}^{75.012}_{75.000}$; the same reasoning applies as for shafts.

Bilateral limits are usually applied when, for example, the position of a hole is specified. The machine operator may position the hole nearer the datum or further from the datum than intended, and, as the operator is in no position to change the situation when the hole has been started, he must aim between the two limits of position, so that the maximum error can be made without causing the part to be rejected. The centre distance between two holes would therefore be specified as, for example, 100 ± 0·02 mm.

(b) Fits

Fits are concerned with the relationship between two parts. Consider a shaft and hole combination: if the shaft is larger than the hole, the condition is said to be of *interference*; and if smaller than the hole, the condition is said to be of *clearance.* The interference may be such that the two parts can be assembled only by shrinking, or it may be very slight, so that the parts can be assembled by hand-operated press. Similarly, the clearance can be slight, so that the shaft can rotate easily in the hole, or be large, so that there is ample clearance for bolts to pass through.

In order that the precise condition is ensured, the limits of size of both the shaft and the hole must be stipulated.

(i) Classes of fit. These are classified in three groups. When the limits of size of both the hole and the shaft are such that the shaft is *always* smaller than the hole, the fit is said to be a *clearance fit.* When the limits of size of both the hole and the shaft are such that the shaft is *always* larger than the hole, the fit is said to be an *interference fit.* When the limits of size of both the hole and the shaft are such that the condition may be of clearance *or* interference, the fit is said to be a *transition fit.*

(ii) Hole-based system and shaft-based system. In order to obtain a range of degrees of clearance, and degrees of interference, it is necessary to use a wide variation of hole sizes and shaft sizes. For example, a manufacturing company could be making a number of parts, all of a nominal 25 mm diameter, but which are all slightly different in actual limits of size, to suit the actual fit required of each pair of parts. This situation could mean that a large number of drills, reamers, gauges, etc. were required.

It is logical that, to reduce this number, a standard hole could be used for each nominal size, and the variation of fit be obtained by making the mating shaft smaller or larger than the hole. This is known as a *hole-based system*. Alternatively, a standard shaft could be used for each nominal size, and the variation of fit be obtained by making the mating hole larger or smaller, as required. This is known as a *shaft-based system*. A hole-based system is usually preferred, because it standardises 'fixed size' equipment such as reamers and plug gauges. A shaft-based system is usually also provided, because sometimes it is more convenient to employ a common shaft to which a number of components are assembled, each with a different fit, and sometimes it is convenient to use bar stock without further machining.

(iii) Systems of limits and fits. It is convenient to establish a standardised system of limits and fits, not only to eliminate the need for the draughtsman to determine the limits each time an assembly is detailed, but also to standardise the tools and gauges required. A system of limits and fits should cater for a wide range of nominal sizes, to satisfy the various needs of industry, and should cater for a wide range of quality of work. The system should, if possible, be tabulated, to save the user the trouble of having to calculate the limits of size to suit the class of fit, the quality of the work, and the size of the part.

(c) British Standard 4500: 1969. ISO limits and fits

This standard replaced BS 1916, which was for both metric and inch sizes. Apart from being completely metric, BS 4500 is essentially a revision of BS 1916 to bring the British Standard into line with the latest recommendations of the International Organisation for Standardisation (ISO). The system refers to holes and shafts, but these terms do not only apply to cylindrical parts but can equally well be applied to the space contained by, or containing, two parallel faces or tangent planes. The system is tabulated, and covers sizes up to 3150 mm.

Figure 13.1 illustrates the principal terms used in this standard. It will be seen that the term *deviation* is used to indicate the position of the limits relative to the *basic size* (nominal size). The basic size is the same for both the hole and the shaft. The zone on

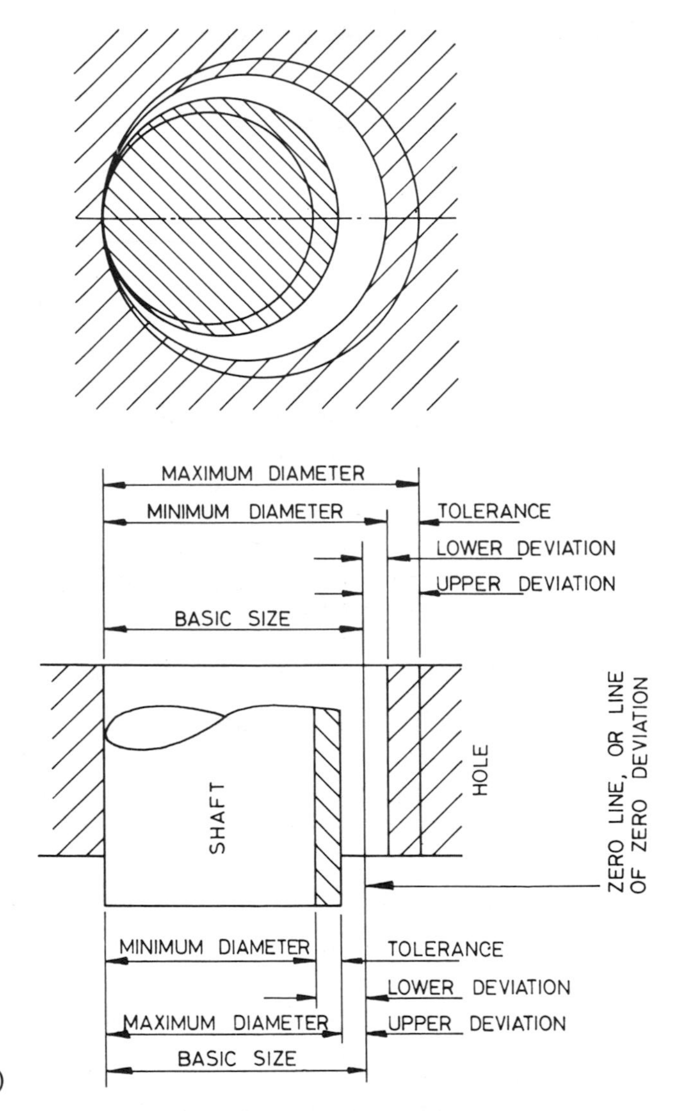

Fig. 13.1 Terms used in BS 4500 (ISO limits and fits)

the diagram between the lines indicating the maximum and minimum diameters is called the *tolerance zone*; it is defined by its magnitude (the *tolerance*) and by its position relative to the zero line. The tolerance zone is useful when illustrating limits and fits, and is used in figs. 13.2, 13.4 and 13.5.

(i) Grades of tolerance. In order to cater for a wide range of quality (grades) of work, there are 18 grades of tolerance in the system, indicated by number: IT 01, IT 0, IT 1, . . . IT 16. (IT stands for ISO series of tolerances, and the higher the number the larger the tolerance for a given size range.) Grades 14 to 16 do not apply for sizes up to and including 1 mm, and grades 01 to 5 do not exist for sizes above 500 mm. Grades 6 to 11 are the commonly used grades. Figure 13.2 compares the magnitude of

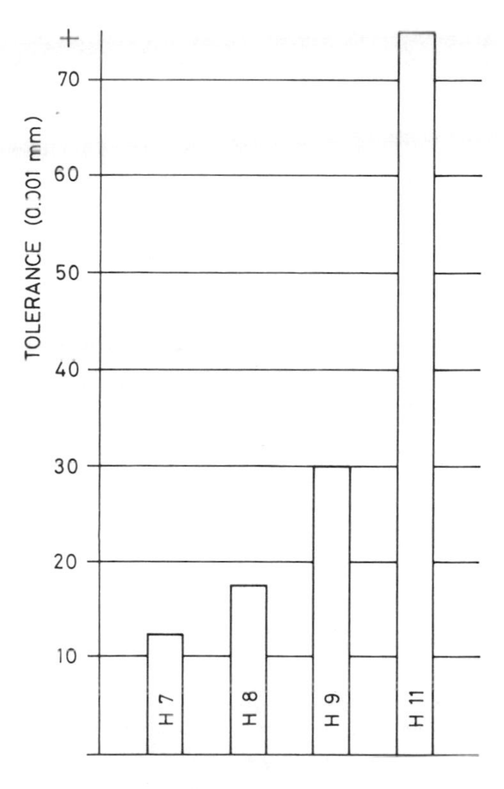

Fig. 13.2 Selected tolerance grades

some of the tolerance grades for the size range 3–6 mm. The H applies to a hole with a lower deviation of zero.

(ii) The effect of size upon tolerance. The tolerance must be increased with the size, because it is more difficult to manufacture and measure a large size than a small size. The connection between tolerance and diameter is illustrated by the curve in fig. 13.3. In order to produce a tabulated system, the increase of tolerance must be in steps (see fig. 13.3 again); to prevent confusion regarding the tolerance to use in the case of diameters where the steps occur, the steps are made in little-used sizes. The

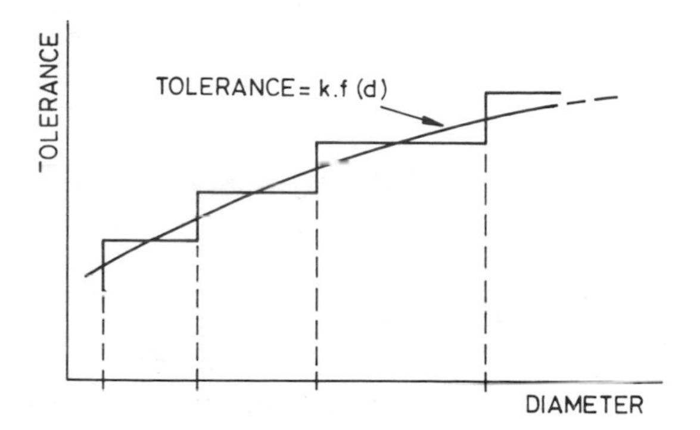

Fig. 13.3 The connection between tolerance and diameter

tolerance is therefore dependent upon the grade of work and the magnitude of the dimension.

(iii) Fundamental deviations. It has already been seen (fig. 13.1) that the *fit* depends not only upon the *tolerance* on the mating parts, but upon the position of the *tolerance zone*. The tolerance zone, relative to the *basic size*, is indicated by the deviation. To provide for a large range of fits, the system includes twenty-seven deviations for holes (indicated by capital letters), and twenty-seven deviations for shafts (indicated by small letters). In order to avoid confusion, the letters i, l, o, q, and w are not used, but some two-letter combinations are used.

In the case of the deviations of holes, the letter indicates the *lower deviation* relative to the zero deviation line (or basic size), the A hole having the greatest positive deviation (i.e. oversized), the H hole having a zero deviation, and the Z, ZA, ZB, and ZC holes having, in turn, the greatest negative deviations (i.e. undersized). The H hole is the one used in the hole-based system. The limits for holes can be designated as follows: A7, H7, ZC6, etc. From this, the lower deviation and the tolerance grade are implied; knowing the size, the actual tolerance, and hence the upper deviation, can be obtained from the tables.

In the case of the deviations of shafts, the letter indicates the *upper deviation* relative to the zero deviation line, the a shaft having the greatest negative deviation (i.e. undersized), the h shaft having a zero deviation, and the z, za, zb, and zc shafts having, in turn, the greatest positive deviations (i.e. oversized). The h shaft is the one used in the shaft-based system. The limits for shafts can be designated as follows: a7, h7, zc6, etc. From this, the upper deviation and the tolerance grade are implied, and, knowing the size, the actual tolerance, and hence the lower deviation, can be obtained from the tables.

In addition to the twenty-seven deviations for holes, and the twenty-seven deviations for shafts, there is a JS hole and a js shaft. In both cases, there is no deviation, but the tolerance is disposed equally about the zero deviation line. These deviations are used when a symmetrical bilateral tolerance is required.

(iv) Tolerance zones. Figs 13.4 and 13.5 illustrate the tolerance zones for some typical fits. Fig. 13.4 illustrates a hole-based system, the hole being an H7 hole, the six fits illustrated being designated as H7-s6, H7-p6, H7-n6, H7-k6, H7-h6, and H7-f6 respectively. The tolerances shown are for the 3–6 mm range.

Fig. 13.5 illustrates a shaft-based system, the shaft being an h6 shaft, and the six fits illustrated being designated as h6-S6, h6-P6, h6-N6, h6-H7, and h6-F7 respectively; the tolerances are for the 3–6 mm range.

(v) BS 4500 and company standards. It will be appreciated that

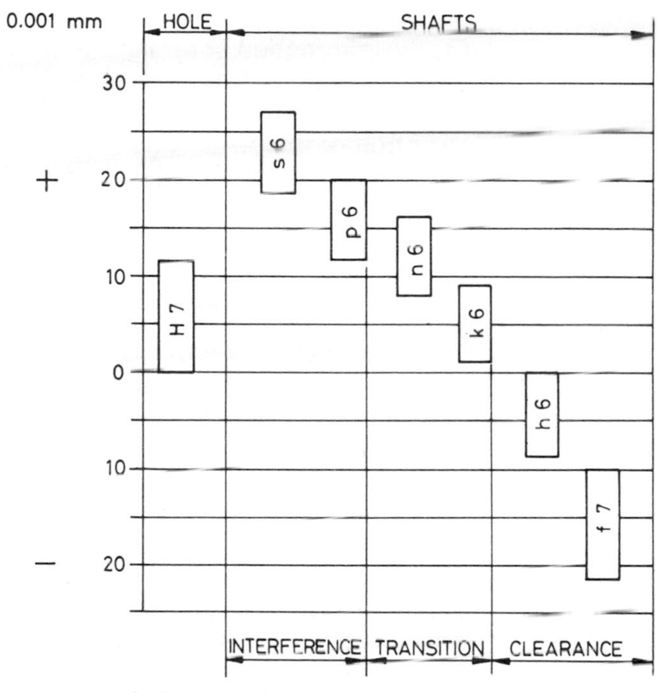

Fig. 13.4 Selected hole-based fits

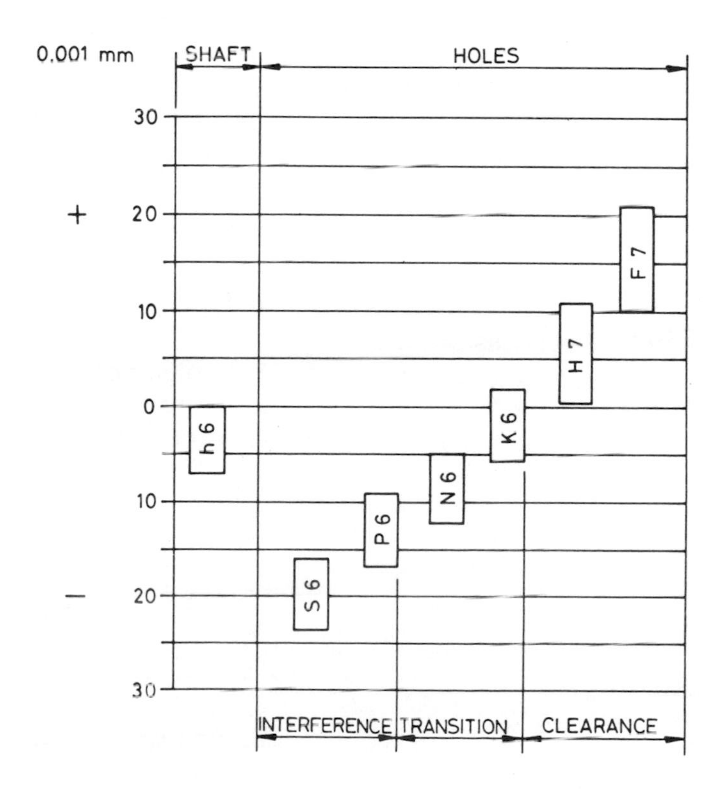

Fig. 13.5 Selected shaft-based fits

BS 4500 covers a very wide range of tolerance grades, sizes, and fits. It is intended that each engineering concern will abstract from the standard the tolerance grades to suit the class of work done, and the size range to suit the size of the products. Again, a selection of the deviations most likely to be required would also be selected, and, from this information, the standard to be used in the work done would be drawn up.

(d) Using BS 4500 to obtain limits

Consider, as an example, the determination of the limits to be applied to a hole and a shaft of 25 mm nominal diameter; the clearance is to be not less than 0.005 mm and no more than 0.045 mm, and a hole-based system is to be used. The upper deviation for the hole is decided first, because a hole is usually more difficult to produce than is a shaft. Tolerance grade IT 7 is considered to be acceptable and the tables in BS 4500 show an H7 to have an upper deviation of 0.021 mm and a lower deviation of zero (an H hole is used because the system is to be hole based). The lower deviation of the shaft must not exceed 0.024 mm (0.045 – 0.021) and its upper deviation must not be less than 0.005 mm (0 – 0.005). A g6 shaft is suitable—the tables in BS 4500 show it to have an upper deviation of 0.007 mm and a lower deviation of 0.021 mm.

The fit would be described on the design scheme as being H7-g6. The detailer would use the tables in BS 4500 to obtain the limits which would be stated on the detail drawings as being 25.000 mm and 25.021 mm for the hole, and 24.993 mm and 24.979 mm for the shaft.

(e) Cumulative tolerances

The accumulation of tolerances must be taken into account when a part is dimensioned. For example, when a part is dimensioned as shown in fig. 13.6, the overall length may vary between 545 and 545.90 mm, i.e. a tolerance of 0.90 mm. It will be seen that the overall tolerance is the sum of the individual tolerances; this accumulation of tolerances can be avoided by dimensioning

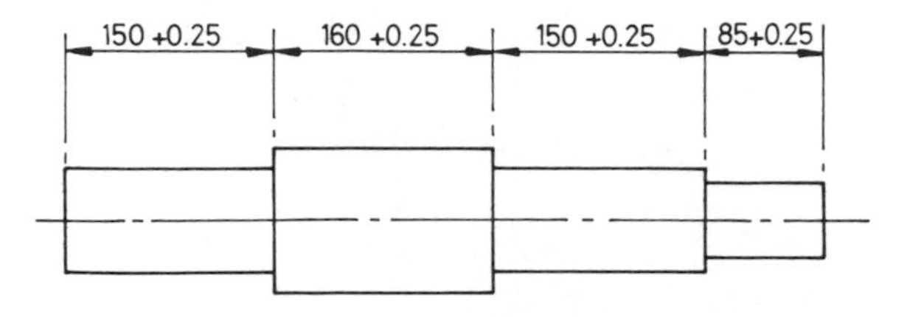

Fig. 13.6 Dimensioning that causes the accumulation of tolerances

each step from one datum, as shown in fig. 13.7. It will be seen that, if one end face was chosen as the datum, it will be difficult to hold the tolerance on the 160 mm length; one end of the 160 mm length is therefore used as the datum. When calculations are made, it is necessary to add the tolerances, even if the dimensions are subtracted; the reader should verify that the tolerance on the 85 mm length is 0.5 mm. If the tolerance on the 85 mm length must be reduced to 0.35 mm, as in fig. 13.6, the tolerances on the 310 and 395 mm lengths must be reduced, so that their sum does not exceed 0.35 mm.

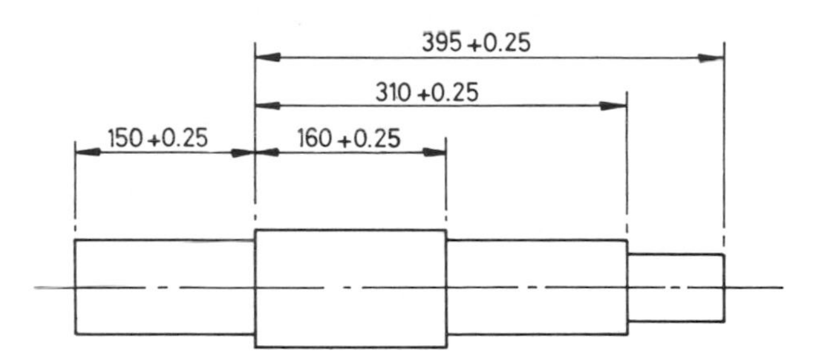

Fig. 13.7 Dimensioning from a datum

The effect of accumulation of tolerances in assemblies needs particular attention. Consider, as an example, fig. 13.8 which illustrates, in block form, an assembly that consists of four discs that are secured to a shaft which, in turn, is located in bearings in a casing and loaded axially so that it is positioned by the head of the bearing shown in fig. 13.8. The maximum and minimum

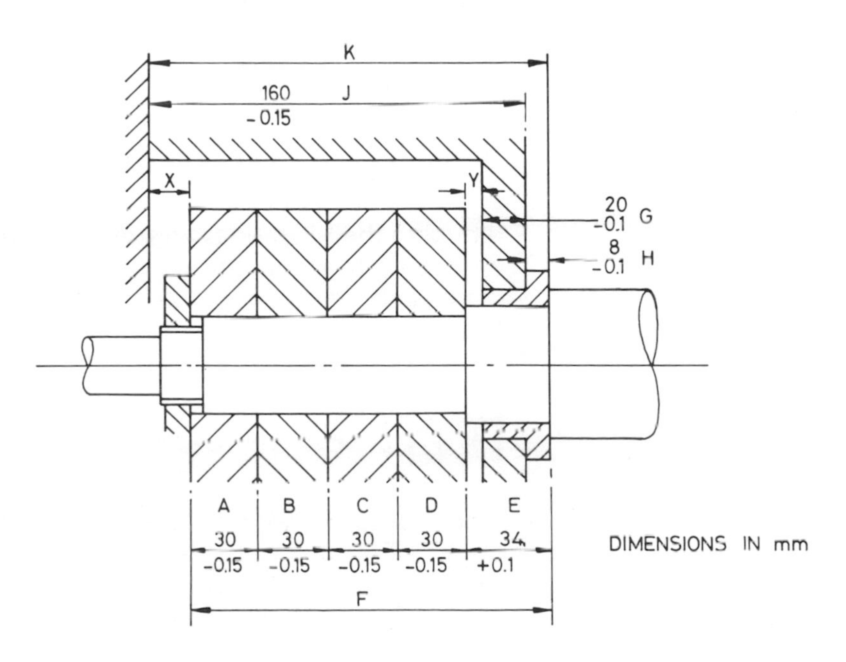

Fig. 13.8 Block diagram representing an assembly

clearance X between the left-hand rotor and the casing cover, and the maximum and minimum clearance Y between the right-hand rotor and the casing are to be determined.

Using the reference letters in fig. 13.8;

$$\text{clearance } X = K - F$$
$$= (J + H) - (A + B + C + D + E).$$

X is a maximum when J and H are at their maximum, and A, B, C, D and E are at their minimum.

$$X_{max} = (160 + 8) - (29.85 + 29.85 + 29.85 + 29.85 + 34) \text{ mm}$$
$$= 168 - 153.4 \text{ mm}$$
$$= 14.6 \text{ mm}$$

X is a minimum when J and H are at their minimum, and A, B, C, D and E are at their maximum.

$$X_{min} = (159.85 + 7.9) - (30 + 30 + 30 + 30 + 34.1) \text{ mm}$$
$$= 167.75 - 154.1 \text{ mm}$$
$$= 13.62 \text{ mm}$$

Clearance $Y = E - (G + H)$.
Y is a maximum when E is a maximum and G and H are at their minimum.

$$Y_{max} = 34.1 - (19.9 + 7.9) \text{ mm}$$
$$= 6.3 \text{ mm}$$

Y is a minimum when E is a minimum and G and H are at their maximum.

$$Y_{min} = 34 - (20 + 8) \text{ mm}$$
$$= 6 \text{ mm}$$

If the tolerances associated with X and Y cannot be held by reducing the tolerances on the individual components, it may be necessary to grind the face of the bush on assembly to a K dimension, or to grind the overall dimension of the rotors $(A + B + C + D)$ when the rotors are assembed on the shaft.

13.2 Geometrical tolerancing

Geometrical tolerancing implies the stipulation of the maximum permissible variation of form or position of a feature and defines the tolerance zone, which is the zone in which the feature is required to be contained.

This section describes the principal features of part three of BS 308 : 1972 Engineering drawing practice, which conforms to the ISO (International Organisation for Standardisation) system of geometrical tolerancing.

(a) Tolerance symbols

The characteristics to be toleranced are classified in BS 308: Part 3 as (i) those of single features and (ii) those of related features.

Tables 13.1, 13.2, 13.3 and 13.4 show the symbols used to indicate the characteristics, the function of geometrical tolerancing as applied to them, a typical example of each characteristic and the tolerance value. (See pages 163–165.)

(b) Tolerance frame

The geometrical tolerance is indicated in a rectangular frame. The frame used for a single feature is shown in fig. 13.9 (this shows that the feature is to be flat to within 0.1 mm) and the frame used for related features is shown in fig. 13.10 (this shows that the feature is to be square with datum A to within 0.1 mm). More than one tolerance frame is used when a feature is to satisfy more than one geometrical tolerance requirement.

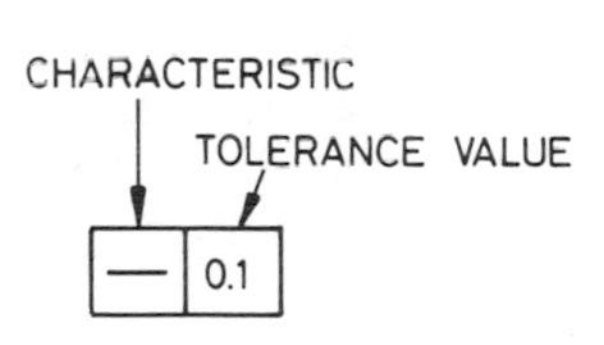

Fig. 13.9 Tolerance frame for a single feature

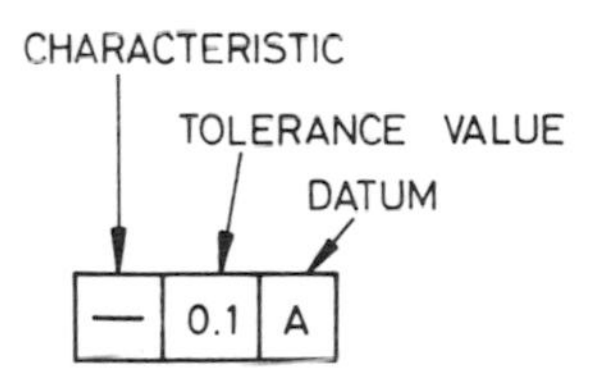

Fig. 13.10 Tolerance frame for related features

(c) Indication of the feature to be controlled

The feature to be controlled is indicated by a leader line from the tolerance frame terminating in an arrow head; the position of the arrow head with respect to the feature is critical because of its implication (see figs 13.11, 13.12 and 13.13).

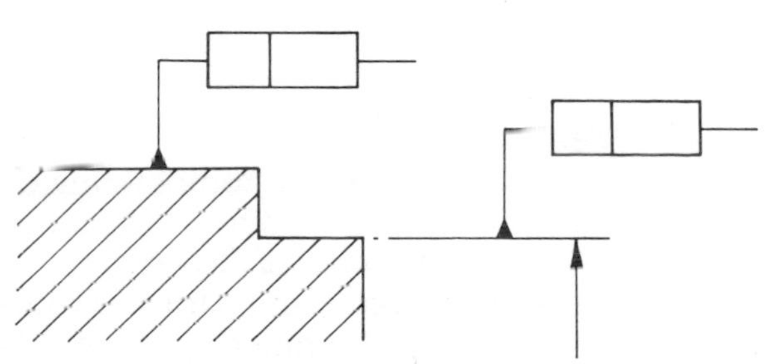

Fig. 13.11 Tolerance frame applied to a line or a surface

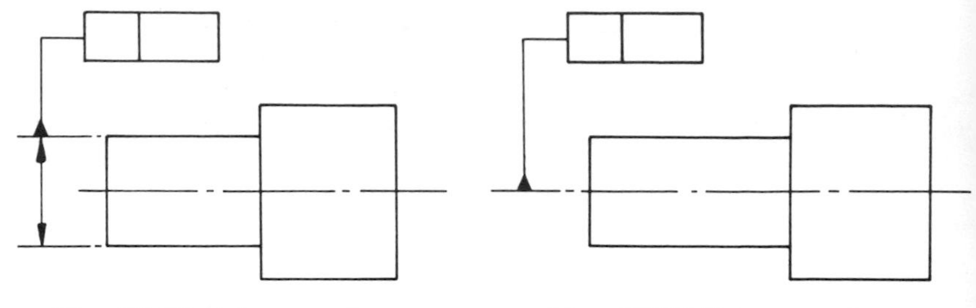

Fig. 13.12 Tolerance frame applied to the axis or median plane of a feature

Fig. 13.13 Tolerance frame applied to common axis or median plane of all features lying on it

(d) Datum systems

Tolerances of parallelism, squareness, angularity, symmetry and of composite tolerance need to be related to a datum feature which is usually subjected to geometrical tolerancing for form. Fig. 13.14 shows two datum features, A and B (these reference letters are boxed), each of which is subject to geometrical tolerancing which would be indicated by the tolerance frames.

A feature may be controlled by more than one datum, in which case more than one datum will be indicated in the tolerance frame. When the datum is a theoretical plane or axis, it is assumed to exist, not in the part itself, but in the precisely made manufacturing or inspection equipment. When the tolerances of position or of profile for a feature are prescribed, the dimensions that define its true position are not toleranced, but are shown instead in a box; this also applies to an angle, when the tolerance of angularity is prescribed.

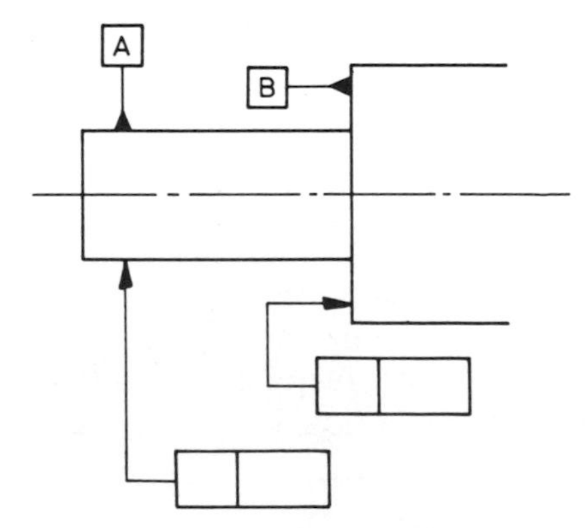

Fig. 13.14 Method of indicating datum features

(e) Maximum material condition

The free assembly of components depends upon the combined effect of actual sizes and geometrical errors. The worst condition for assembly occurs when the mating components are at maximum material limits for size together with maximum permitted geometrical errors. If a component is of a size that is other than its maximum material limit there will be an increased

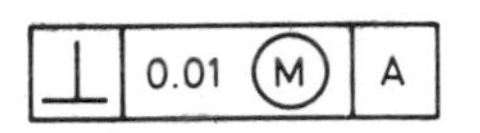

Fig. 13.15 Indicating relaxation of geometrical tolerance

clearance which will permit assembly even if the geometrical tolerance is slightly relaxed.

When it is possible for the geometrical tolerance to be relaxed it is indicated by Ⓜ in the tolerance frame as shown in fig. 13.15.

Table 13.1 Tolerance of form of single features

Characteristic and symbol	Function of geometrical tolerance	Tolerance zone
Straightness	To control the straightness of a line on a surface, or an axis of a feature.	Area between two parallel straight lines in the plane containing the considered line or axis. Tolerance value is the distance between them. 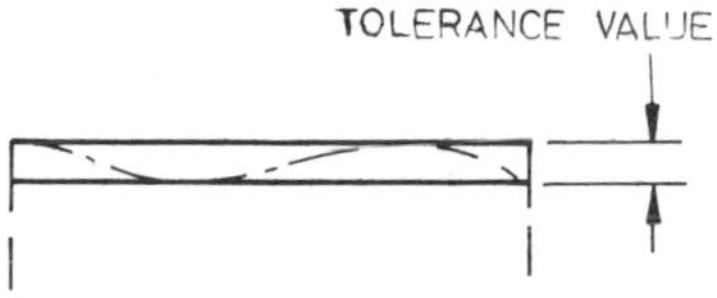
Flatness	To control the flatness of a surface.	Area between two parallel planes. Tolerance value is the distance between them. 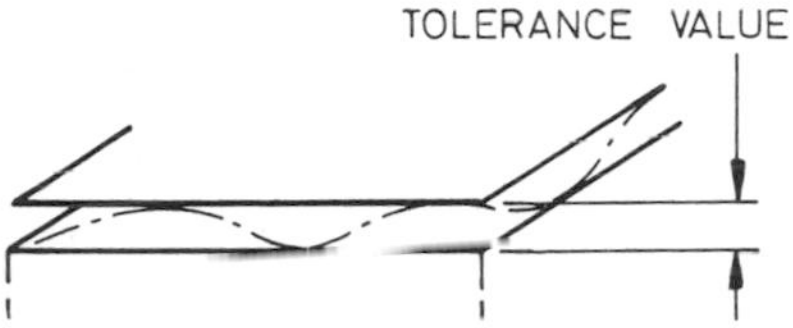
Roundness	To control the errors of form of a circle in the plane in which it lies. The circle may be a section of a solid of revolution. It is not concerned with the position of the circle.	Area between two concentric circles. Tolerance value is the radial distance between the circles. 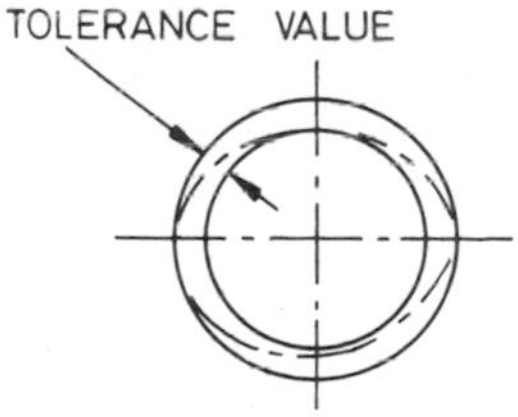
Cylindricity	To control a combination of roundness, straightness and parallelism applied to the surface of a cylinder.	Annular space between two cylinders that are co-axial with each other. Tolerance value is the radial distance between them. 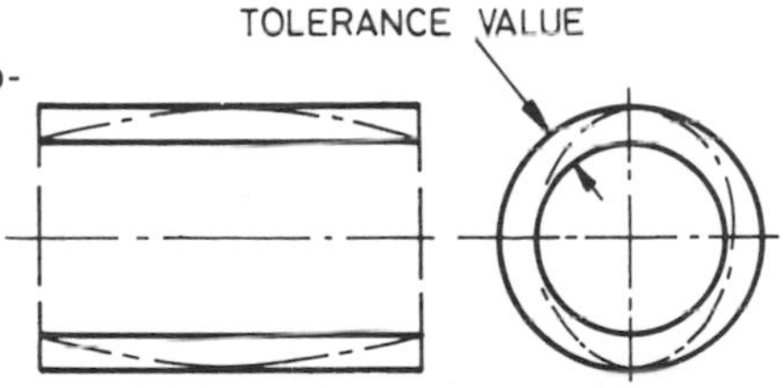
Profile of a line	To control the shape of a profile.	Area between two lines which envelop circles of diameter equal to the tolerance zone. Either the centre line of the circles or one line lies on the perfect profile. 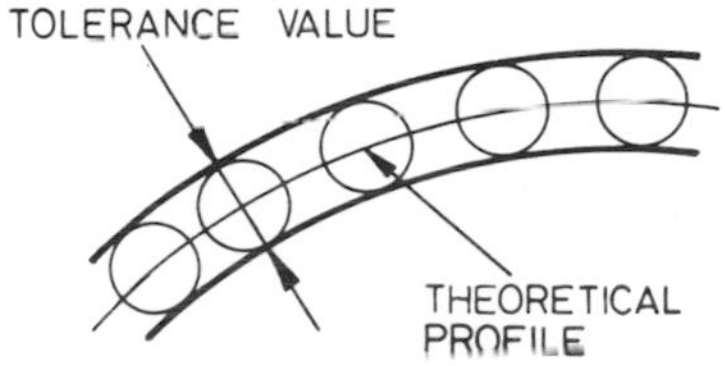
Profile of a surface	To control the shape of a profile.	Area between two surfaces which envelop a series of spheres with their centres on the surface having the correct geometrical shape. 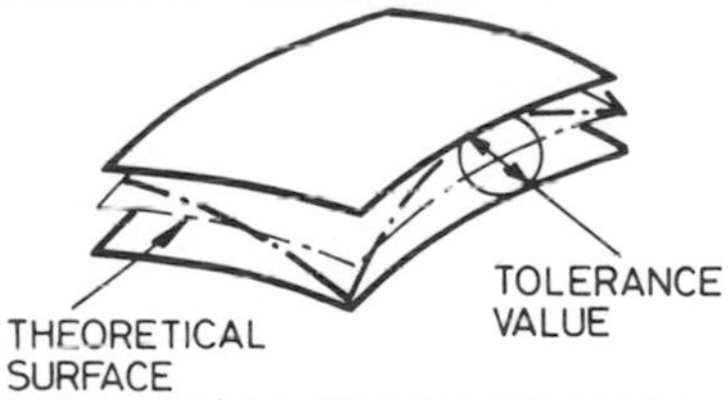

Table 13.2 Tolerance of attitude of related surfaces

Characteristic and symbol	Function of geometrical tolerance	Tolerance zone
Parallelism //	To control parallelism of a line or surface with respect to a datum feature which may be a line or a plane. Applied to parallelism of axes of holes, axis of a hole and surface or to two surfaces or lines.	Area between two parallel lines or space between two parallel lines which are parallel to the datum feature. Tolerance value is the distance between the two lines or planes. TOLERANCE VALUE DATUM 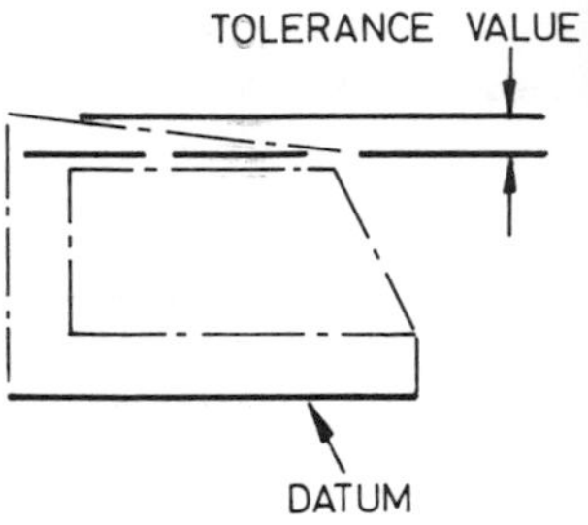
Squareness ⊥	To control squareness of a line or surface with respect to a datum feature which may be a line or a plane. Applied to the axes of holes, axis of a hole and a plane surface or to two surfaces of lines.	Area between two parallel lines or space between two parallel planes which are perpendicular to the datum feature. Tolerance value is the distance between the two lines or planes. TOLERANCE VALUE DATUM 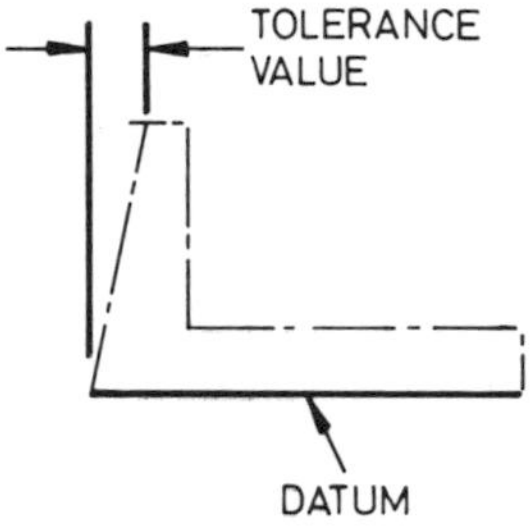
Angularity ∠	To control inclination of a line or surface with respect to a datum feature which may be a line or a plane. Applied to axes of two holes, axis of a hole and a plane, two lines or two planes.	Area between two parallel lines or space between two parallel planes which are inclined at a specified angle to the datum feature. Tolerance value is the distance between the two lines of planes. TOLERANCE VALUE θ DATUM 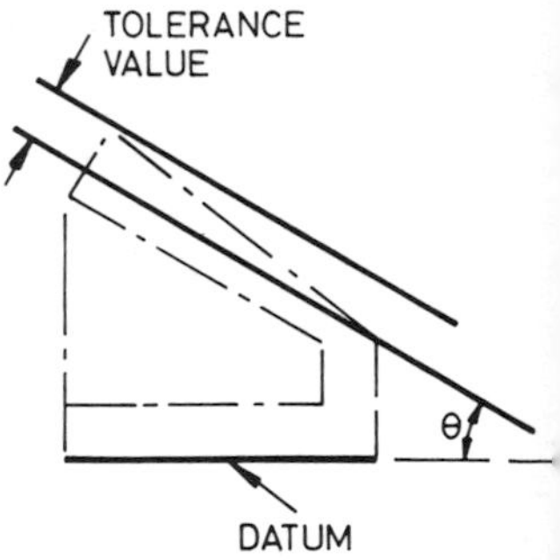

Table 13.3 Tolerance of location of related surfaces

Characteristic and symbol	Function of geometrical tolerance	Tolerance zone
Position ⌖	To control the deviation of the position of a feature from its specified true position.	Circle in which the centre of a hole must lie, cylinder in which the axis of a hole must be contained, area between two straight and parallel lines which are symmetrically disposed about the specified true position of the line.
Concentricity ◎	Applied to circles or cylinders to control the deviation of the position of the centre or axis of the toleranced feature relative to the centre or axis of the datum feature.	Centre or axis to lie within circle or cylinder. Tolerance value is the diameter of such a circle or cylinder. 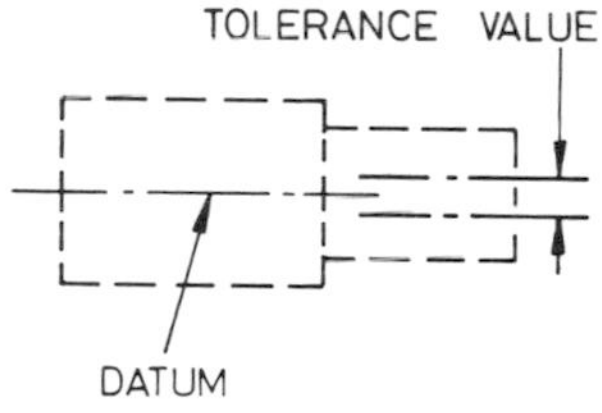 TOLERANCE VALUE DATUM
Symmetry ⌯	To control the deviation of the position of a feature whose position is specified by its symmetry relative to a datum. Applied to lines, slots etc.	Area between two parallel lines or planes that are symmetrically disposed about a datum line. Tolerance value is the distance between them. 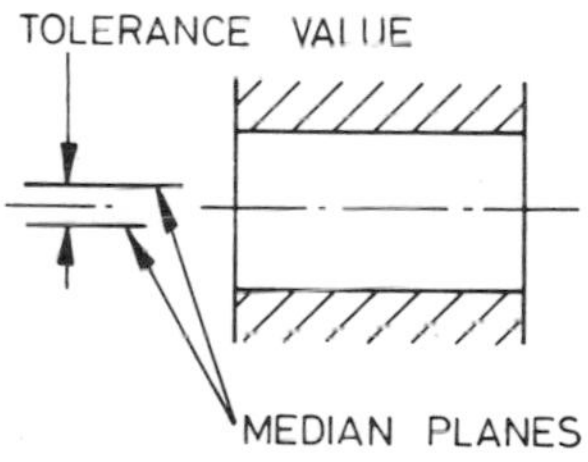 TOLERANCE VALUE MEDIAN PLANES

Table 13.4 Composite tolerance of related features

Characteristic and symbol	Description	Tolerance
Run-out ↗	Defined in terms of measurement under rotation. Related to cylinder, cone or end face with respect to axis of rotation.	Maximum permissible variation of position (full indicator movement) of the considered feature with respect to one complete revolution about the datum axis without axial movement.

14 Specification of Surface Texture

The surface of parts produced by casting, working, and powder metallurgy will have irregularities associated with the manufacture of the dies etc. used in their production, and with the production process itself; irregularities will still be present after the parts have been machined.

Surface irregularities are of three types, as illustrated by fig. 14.1 which shows the profile of a nominally flat surface.

There is a general error, known as *geometric* (*or form*) *error* (illustrated separately in fig. 14.1 (a). This error can easily be detected by a straight edge, dial indicator etc.; it is caused by faulty machine alignment, faulty machine slides, a worn chuck or cutter spindle, worn bearings, incorrect machine setting and so on. It is controlled by geometrical tolerancing as discussed in the previous chapter. Superimposed upon this error (or upon an otherwise 'true' surface) there are two further sets of irregularities (shown separately in fig. 14.1 (b)) that together tend to form a

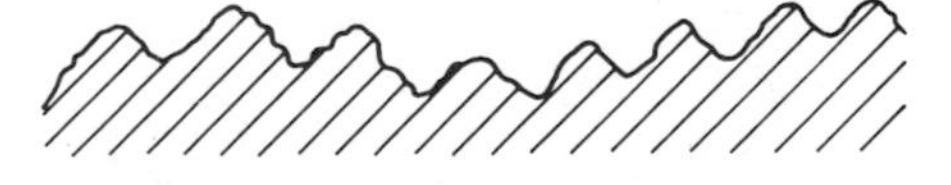

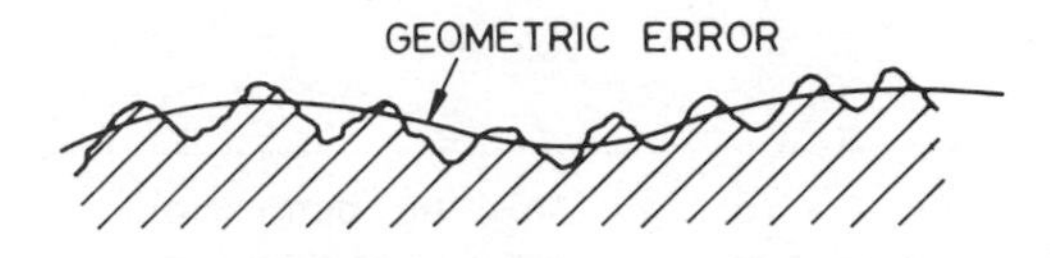

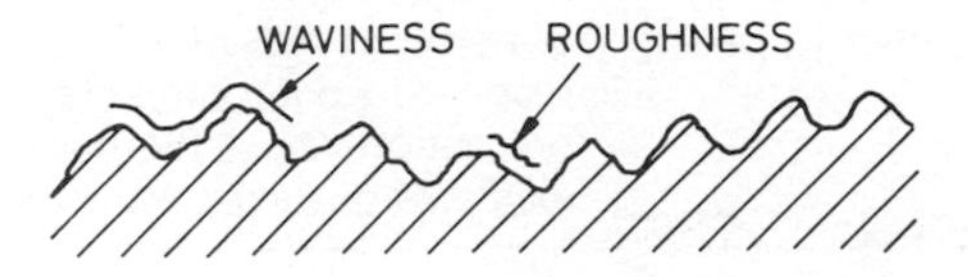

Fig. 14.1 Surface irregularities; (a) geometric error; (b) waviness and roughness

pattern or texture on the surface; this pattern is known as *surface texture*. The larger of these irregularities is known as *waviness*, and is caused by spindle or cutter deflection, machine vibrations, chatter, warping strains etc. Waviness has, in turn, irregularities superimposed on it known as *roughness*; roughness can be defined as 'irregularities in the surface texture which are inherent in the production process, but excluding waviness and form errors'.

The principal reasons for controlling surface texture are:

- to reduce the initial wear of parts that are in contact (the 'running in');
- to improve the fatigue resistance (surface irregularities are the seat of fatigue failure);
- to allow fine geometric and dimensional tolerances to be held (this cannot be done if the surfaces involved have irregularities that are nearly as large as the tolerance allowed on their positions);
- to reduce frictional wear (smooth contacting surfaces will wear less rapidly than rough surfaces);
- to reduce corrosion by minimising the number and depth of crevices, where corrosion proceeds at a high rate.

This does not imply that a perfectly smooth surface is always ideal; for example, a controlled degree of roughness is usually required in a cylinder bore to allow 'reservoirs' of oil to be present.

(a) The assessment of surface texture

The assessment of surface texture quality is the subject of BS 1134:1972. Fig. 14.2 illustrates some of the terms used in conjunction with the study of surface texture.

Surface texture can be examined by optical methods, tactual methods or by methods that use a stylus. The latter can be used to produce a trace which gives a picture of the profile of the surface and a numerical evaluation of the texture. A numerical value does not give a complete assessment of texture but it is useful when a known process is to be controlled, and is therefore applicable to production drawings and other instructions.

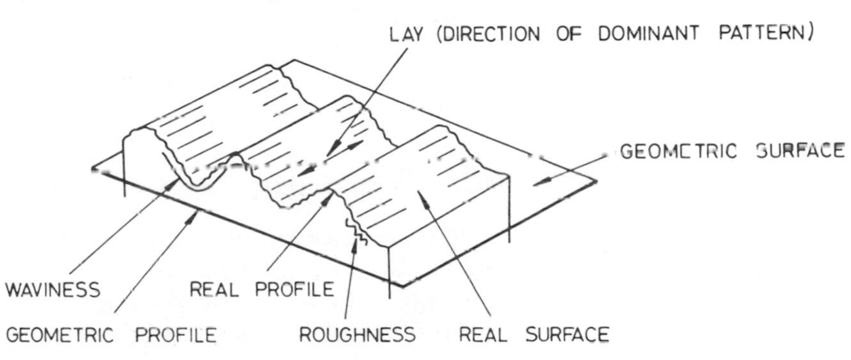

Fig. 14.2 Illustration of some terms used in the study of surface texture

(b) The numerical evaluation of surface texture

The most commonly used numerical roughness value is the arithmetic average deviation of the surface from a datum; this is obtained by passing a stylus across the surface in a direction approximately at right angles to the lay if the surface texture has a directional quality. The value so obtained is termed the R_a value; it can be evaluated from a trace or directly by the instrument. The R_a value is usually quoted in μm (1 μm = 0.000001 m) and its specified value should be selected from the preferred values shown in table 14.1 which also shows the corresponding μ in values (1 μ in = 0.000001 in). A series of roughness grade numbers extracted from International Standard ISO/R 1302 (Technical drawings; methods of indicating surface texture on drawings) which may be used on drawings to avoid the misinterpretation of numerical values are also shown in table 14.1.

Table 14.1 Preferred R_a values

Nominal R_a values		Roughness grade number
μm	μ in	
50	2000	N 12
25	1000	N 11
12.5	500	N 10
6.3	250	N 9
3.2	125	N 8
1.6	63	N 7
0.8	32	N 6
0.4	16	N 5
0.2	8	N 4
0.1	4	N 3
0.05	2	N 2
0.025	1	N 1
0.0125	0.5	—

Table 14.2 shows typical surface roughness values produced by common production processes.

The length over which the study is made will affect the value of R_a; if the waviness is to be fully included, the examination must be over at least one waviness wavelength. The length over which the study is made is called the *sampling length* and is standardised by BS 1134 as:

0.08	0.25	0.8	2.5	8.0	25.0	millimetres
0.003	0.01	0.03	0.1	0.3	1.0	inch

Electrical instruments are designed to traverse the surface so that several sampling lengths are included. The recording meter cuts off at the end of each sampling length (called the *meter cut-off length*) and produces an average value for that length; the

Table 14.2 Surface roughness values produced by typical common production processes

(After BS 1134)

Key: Average application (solid bar); Less frequent application (hatched bar)

Roughness values (μm R_a): 50, 25, 12.5, 6.3, 3.2, 1.6, 0.8, 0.4, 0.2, 0.1, 0.05, 0.025, 0.0125

Process	Less frequent application	Average application	Less frequent application
Sand casting	50–25	25–12.5	12.5–6.3
Hot rolling	50–25	25–12.5	12.5–6.3
Forging	25–12.5	12.5–3.2	3.2–1.6
Permanent mould casting	6.3–3.2	3.2–1.6	1.6–0.8
Investment casting	6.3–3.2	3.2–0.8	0.8–0.4
Extruding	12.5–3.2	3.2–0.8	0.8–0.4
Cold rolling, drawing	6.3–3.2	3.2–0.8	0.8–0.2
Die casting	3.2–1.6	1.6–0.8	0.8–0.4
Flame cutting	50–25	25–12.5	12.5–6.3
Sawing	25–12.5	12.5–3.2	3.2–1.6
Planning, shaping	25–12.5	12.5–0.8	0.8–0.4
Drilling	12.5–6.3	6.3–1.6	1.6–0.8
Chemical milling	12.5–6.3	6.3–1.6	1.6–0.8
Electro-discharge machining	12.5 to between 6.3 and 3.2	between 6.3 and 3.2 to 1.6	1.6–0.8
Milling	25–6.3	6.3–0.8	0.8–0.2
Reaming, broaching	6.3–3.2	3.2–0.8	0.8–0.4
Grinding	6.3–1.6	1.6–0.1	0.1–0.025
Honing	1.6–0.8	0.8–0.1	0.1–0.025
Lapping	0.8–0.4	0.4–0.05	0.05–0.0125

final value displayed is the average for all such values. The cut-off length is made equal to the corresponding sampling length, and is related to the finishing process used for the surface being examined (see table 14.3)

Many electrical instruments include a filtering device that can be set to exclude large irregularities from the recording. This is

Table 14.3 Process designations and suitable cut-off values (after BS 1134)

Typical finishing process	Designation	Meter cut-off (mm)				
		0.25	0.8	2.5	8.0	2.5
Milling	Mill		■	■	■	
Boring	Bore		■	■	■	
Turning	Turn		■	■		
Grinding	Grind	■	■	■		
Planing	Plane			■	■	■
Reaming	Ream		■	■		
Broaching	Broach		■	■		
Diamond turning	D. Turn	■	■			
Diamond boring	D. Bore	■	■			
Honing	Hone	■	■			
Lapping	Lap	■	■			
Shaping	Shape		■	■	■	
Electro-discharge machining	EDM	■	■			
Drawing	Draw		■	■		
Extruding	Extrude		■	■		

important when errors caused by the machine must be excluded so that the process itself can be studied.

(c) The specification of surface texture

BS 1134 states that the surface roughness value should be expressed in numbers of μm (alternatively by N number). For example: 0.4 μm R_a. This implies that the value accepted is between zero and 0.4 μm and that, as the sampling length is not stated, it is 0.8 mm. If minimum and maximum values are to be specified, they should be expressed as:

$$\begin{matrix} 0.4 \\ 0.8 \end{matrix}\ \mu R_a \text{ or } 0.4\text{–}0.8\ \mu\text{m } R_a.$$

Should the sampling length be other than 0.8 mm, that value should be indicated in parenthesis following the surface roughness value. For example: 0.2 μm R_a (2.5). The specification (either directly or by implication) of the sampling length to be used in the assessment, also denotes that the spacing of the dominant peaks must not exceed that value.

When the direction of the lay affects the performance of the component, it must be specified following that of roughness and, where applicable, that of the sampling length. For example: 6.3 μm R_a lay circular. BS 1134 gives six basic lays and appropriate symbols to use on drawings. These are: straight lay along the surface, straight lay across the surface, crossed lay, approximately circular lay relative to the centre of the surface, approximately radial lay relative to the centre of the surface, and multi-directional.

(d) Surface texture and drawing symbols

As stated in BS 308: Part 2 Dimensioning and tolerancing of size, surface texture can be specified using machining symbols. It is necessary to state the roughness value, but the sampling length may either be stated, or implied to be 0.8 mm by its omission. The production method (i.e. hone, grind) is stated if a particular method is required. Similarly, the direction of the lay can, if required, be stated by a note or, alternatively, by a symbol (see table14.4).

The location of the surface texture details when used in conjunction with a machining symbol is shown in fig. 14.3.

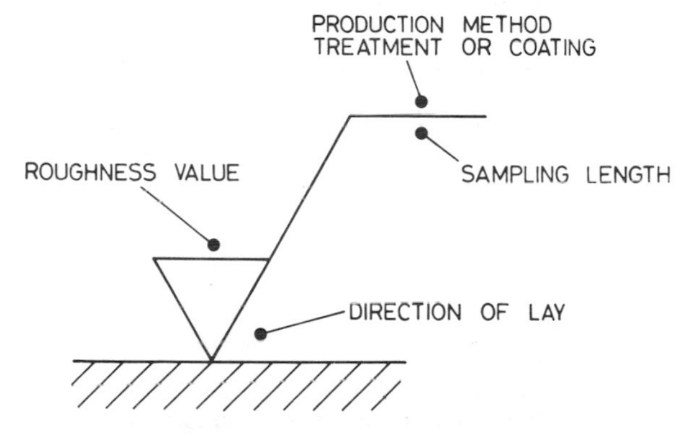

Fig. 14.3 Location of the surface texture details when used with machining symbol

When the surface texture details refer to a machined surface, the symbol shown in fig. 14.4 would be used. When it is immaterial whether or not the surface is machined, a symbol of the form

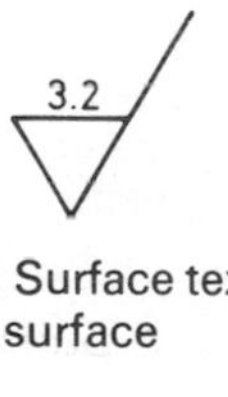

Fig. 14.4 Surface texture and machined surface

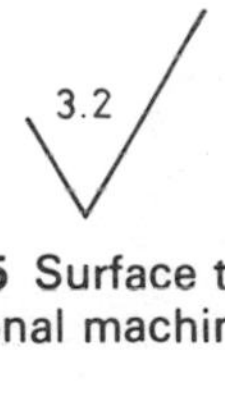

Fig. 14.5 Surface texture and optional machining

Fig. 14.6 Surface texture indication when machining not allowed

shown in fig. 14.5 should be used. Where it is necessary to indicate surface texture details and the surface must not be machined, a note to that effect can be added or, alternatively, a symbol of the form shown in fig. 14.6 can be used.

Table 14.4 Symbols of the direction of lay (after BS 308: Part 2)

Lay symbols	Explanation and application
=	Parallel to the plane of projection of the view in which the symbol is used. 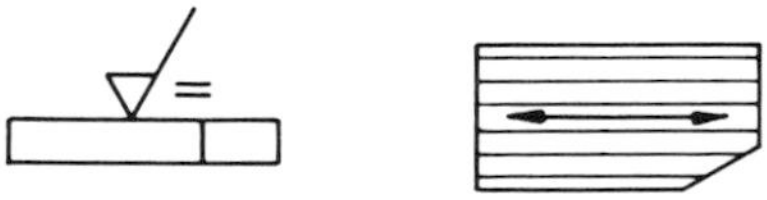
⊥	Perpendicular to the plane of projection of the view in which the symbol is used. 
X	Crossed in two slant directions with regard to the plane of projection of the view in which the symbol is used. 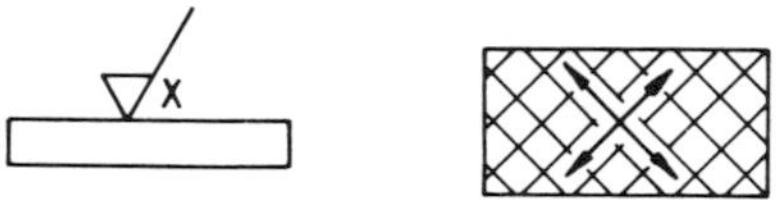
M	Multi-directional. 
C	Approximately circular relative to the centre of the surface to which the symbol is applied.
R	Approximately radial relative to the centre of the surface to which the symbol is applied.

Assignments

Certain of these assignments are suitable for group projects or class discussion.

1 When the requirements of a proposed device or product are considered, it is usually necessary to consider the functions under primary functions, and secondary functions—the latter often limit the extent to which the primary functions can be satisfied. Identify the primary functions and the secondary functions in each of the following pairs, and explain how the satisfaction of the secondary functions affects the design specification.

(a) An easy chair for use in the home and a chair for use in a community hall which is used on some evenings for meetings and on others for sports activities etc.

(b) A camera for professional studio work and a camera for a novice.

(c) A motor car for family use and a racing car.

2 A grass cutter is required to cut the grass that forms the paths on an allotment site. The site does not have access for cars, and there is no mains electricity supply available. List the basic requirements of the cutter and the methods that could be used to power it. Use a rating system to identify the power system that best satisfies the basic requirements.

3 Examine a typical product, list all its component parts and state both the function of each part and the manufacturing requirements of the material from which each is made. Identify the material from which each part is made and relate its characteristics to the service and manufacturing requirements.

4 Given a pictorial or arrangement drawing of a simple assembly, identify the principal component parts and their service and manufacturing requirements. Suggest the material from which each might be made.

5 Consider a specific component and list its service requirements and those of the ideal manufacturing method that most nearly satisfies the requirements. Examine the actual component to identify the manufacturing method used, and compare its characteristics with those of the ideal manufacturing method.

6 Examine some typical components to identify the manufacturing method used. Suggest the factors that the designer took into account when selecting the manufacturing method, and the reasons for his choice.

7 Given a basic design scheme, analyse the circumstances (for example, size, cost, quantity and material) and specify the casting method to be used for appropriate components.

8 Given a basic design scheme, produce a detail drawing of the casting(s) involved to suit (a) sand-casting, (b) die-casting and (c) investment-casting.

9 Given the detail drawing of a casting designed for a specific casting process, redesign the casting so that a different casting method can be used.

10 The manufacture of a metal gear may include the following processes:— (i) casting, (ii) forging, (iii) powder metallurgy and (iv) presswork.
(a) List the advantages and limitations of each of the four processes as applied to the manufacture of gears, and name a product for which each process could be employed to advantage.
(b) Produce drawings of gears to show the shape produced by each of the four primary processes, and also the extent of the work that is necessary to complete the gear.

11 Examine an aluminium alloy extending ladder; sketch the sections of the extruded material used in its construction and comment upon the effectiveness of this form of construction.

12 Given a drawing of a component produced by a method such as casting or machining from bar, redesign the component so that powder metallurgy can be used instead.

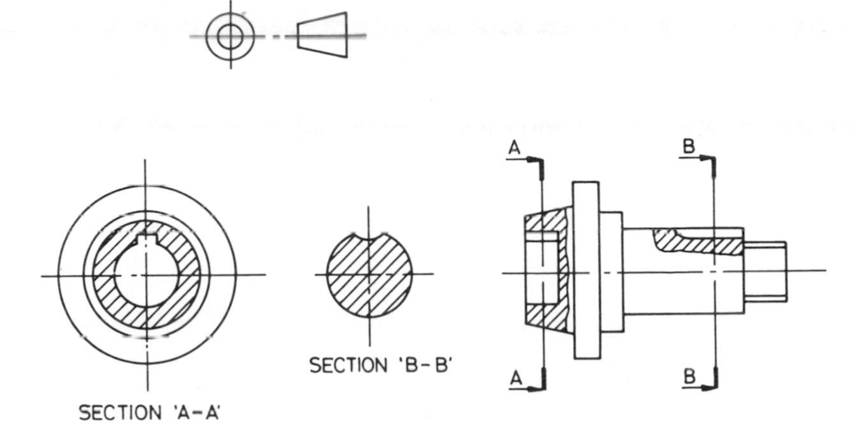

Fig. A13 Spindle

13 Fig. A13 shows, half-size, a component that is to be machined all over. Identify and correct the design faults.

14 Examine several products in which joining and fastening processes have been used. Comment upon the advantages, and possible disadvantages, of the joining or fastening method used in each product.

15 Study domestic products, a motor car or a cycle in which corrosion has occurred. Suggest methods whereby corrosion could be reduced and relate the methods to the cost of the product.

16 Plastics materials can be used instead of metals. Examine several products that include plastics components and comment upon the effectiveness of using plastics.

17 Study, and appraise, the ergonomics of typical equipment or products. Examples might include:
(a) the control arrangement and display in a motor car;
(b) a vacuum cleaner;
(c) a refrigerator or freezer;
(d) a music centre;
(e) a camera;
(f) the layout of a working area such as a kitchen or office.

18 Design a clothes-line system such that in the event of a storm, the clothes, whilst remaining on the line, can be 'gathered up' into a shelter and be moved out, to continue to dry, when the storm has passed.

19 Design a system whereby the top of a typical private motor car can be used as a grandstand at horse trials, point-to-point races, and similar events. The grandstand must be able to carry two adults and two children with safety. It is intended that it will be positioned on the car at home, and the final assembly done very quickly when required.

20 Design an alignment attachment for use with a handyman's hand-held drill to ensure that the drilled hole is square with the plane of the workpiece surface.

21 Corner-section kitchen units tend to be awkward to use and to clean. Examine this problem and design a corner unit that makes better use of the storage space.

22 Heavy suitcases can be more easily moved using wheels. Consider the ways whereby wheels can be introduced and, by using a rating system, identify the best system. Produce a design scheme showing the principal features of the system that you have selected.

23 Design a small hand-operated rolling mill for use on thin metal. The mill should be adjustable to suit a range of initial thicknesses, and to allow the thickness to be reduced in stages so that the effect of the rolling on the properties and structure of the metal can be studied.

24 Design a small paper-folding machine for use by clubs, churches and similar societies when producing programmes etc. on a small budget. The object of using the folding machine is to make the job less tedious and more accurate than hand-folding, and not to increase the speed of the operation.

Index